Theory and Applications of Stochastic Processes

Theory and Applications of Stochastic Processes

Editor

Jacek Fabian

scitus academics

Theory and Applications of Stochastic Processes
Edited by **Jacek Fabian**

ISBN: 978-1-68117-652-9
Library of Congress Control Number: 2016946519

www.scitusacademics.com

Printed in United States of America on Acid Free Paper

Preface

Everyday we encounter signals which cannot be modeled exactly by an analytic expression or in a deterministic way. Examples of such signals are ordinary speech waveforms, seismological signals, biological signals, temperature histories, communication signals etc. In manufacturing domain no machine is totally reliable. Every machine fails at some random time. Thus in a typical manufacturing system which involves a large number of machines, the total number of machines at any time cannot be determined in a deterministic way. In a market driven economy, the stock market is volatile, the interest rates fluctuate in a random fashion. One can give any number of examples from our daily life events where uncertainty prevails in an essential way. This gives us the realization that many real life phenomena require the analysis of a system in a probabilistic setting rather than in a deterministic setting. Thus stochastic models are becoming increasingly important for understanding or making performance evaluation of complex systems in a broad spectrum of fields. A stochastic process is simply a collection of random variables indexed by time. Stochastic Analysis deals with models which involve uncertainties or randomness. Uncertainty, complexity and dynamism have been continuing challenges to our understanding and control of our physical environment.

Theory and Applications of Stochastic Processes presents an analytical approach to stochastic processes. Its aim is to make probability theory readily accessible to scientists trained in the traditional methods of applied mathematics, such as integral, ordinary, and partial differential equations and in asymptotic methods, rather than in probability and measure theory. It will appeal to advanced graduate students, researchers and practitioners in mathematics, physics and engineering.

Stochastic processes have played a significant role in various engineering disciplines like power systems, robotics, automotive technology, signal processing, manufacturing systems, semiconductor manufacturing, communication networks, wireless networks etc. To realize the above goal of building prosthetic limbs, one tool which plays a critical role is the theory of stochastic processes.

Table of Contents

CHAPTER 1

Stochastic Process Optimization Technique

Hiroaki Yoshida[1], Katsuhito Yamaguchi[2], Yoshio Ishikawa[1]

[1] College of Science and Technology, Nihon University, Chiba, Japan

[2] Junior College, Nihon University, Chiba, Japan

ABSTRACT

The conventional optimization methods were generally based on a deterministic approach, since their purpose is to find out an accurate solution. However, when the solution space is extremely narrowed as a result of setting many inequality constraints, an ingenious scheme based on experience may be needed. Similarly, parameters must be adjusted with solution search algorithms when nonlinearity of the problem is strong, because the risk of falling into local solution is high. Thus, we here propose a new method in which the optimization problem is replaced with stochastic process based on path integral techniques used in quantum mechanics and an approximate value of optimal solution is calculated as an expected value instead of accurate value. It was checked through some optimization problems that this method using stochastic process is effective. We call this new optimization method "stochastic process optimization technique (SPOT)". It is expected that this method will enable efficient optimization by avoiding the above difficulties. In this report, a new optimization method based on a stochastic process is formulated, and several calculation examples are shown to prove its effectiveness as a method to obtain approximate solution for optimization problems.

KEYWORDS

Optimization, Stochastic Process, Path Integral, Expected Value, Quantum Mechanics

1. INTRODUCTION

Optimization methods are useful to system design, and especially the applications to engineering problems are expected. In order to obtain the accurate optimal solution, generally the conventional optimization methods were based on a deterministic approach. However, when design variables have many inequality restriction conditions, in order to search for a solution efficiently, it is necessary to adjust a parameter with trial and error. Moreover, if the problem is complex, the risk of falling into a local solution is also high.

Our method obtains an approximate value of an optimal solution using stochastic process. The optimization methods using stochastic process like simulated annealing (SA) [1] are found. However, in these methods, stochastic process is used as a means of searching for an accurate solution efficiently. Our method is based on probability theory, and essentially differs from the concepts of the conventional optimization methods. The purpose of the conventional optimization is to obtain the accurate optimal solution, but that of our method is to obtain the approximate value of the optimal solution.

We showed that the approximate value of the optimal solution could be obtained using stochastic process [2] [3] . And even if it was an approximate value of the optimal solution obtained by this method, it was shown that it is useful in engineering [4] . Furthermore, it was shown that this method can apply to general engineering optimization problems containing static design variables and dynamic design variables [5] .

As mentioned above, it was shown that the approximate value of the optimal solution can be obtained by this method using stochastic process. The aim of this paper is not to show the results of the applications of our method, but to bring our original idea to a wide reading audience.

Thus, we propose a new method in which an optimization problem is replaced with a stochastic process based on path integral techniques used in quantum mechanics and an approximate value of an optimal solution is calculated as an expected value instead of an accurate value. We call this new optimization method the "stochastic process optimization technique (SPOT)". It is expected that this method will enable efficient optimization by avoiding the above difficulties.

Since expected values are stochastic average values, a stochastic approach is not appropriate to obtain an accurate solution as a deterministic approach. However, its characteristics are simple algorithms and the fewer number of manual parameters. The resultant solution is approximate enough to a global solution of engineering problems whose solution spaces are expected to be relatively smooth. The approximate solution can be an initial value to obtain an accurate solution with a deterministic process.

In this paper, a new optimization method is formulated based on a stochastic process. Some calculation examples are introduced to show how effective the approximate solutions are for optimization problems. The examples are a simple benchmark test, a classic line of swiftest descent problem, and an optimized glide problem of a hang glider as an introductory engineering optimization problem.

2. PATH INTEGRAL AND OPTIMIZATION

According to the principle of least action, Newtonian motion of a mass point makes the action integral minimum:

$$I = \int_{t_1}^{t_2} L(x_i, \mathrm{d}x_i/\mathrm{d}t, t)\,\mathrm{d}t \quad (1)$$

where $L(x_i, \mathrm{d}x_i/\mathrm{d}t, t)$ is the Lagrange function. Also, the Euler-Lagrange equation obtained with variation of I is equal to an equation of Newtonian motion. On the other hand, the motion of a particle is determined stochastically according to the quantum mechanics. A motion on the classical mechanics, which keeps I minimum, occurs at the "highest possibility". However, other motions can occur at certain probabilities. This results in some particular phenomena, such as the tunnel effect, which never occur on the classical mechanics.

The well-known simulated annealing (SA) method adopts such a quantum mechanical approach to the solution of optimization problems. SA introduces a probability distribution that reaches its maximum at an optimal value. This method numerically searches the optimal value (peak of the probability distribution) with a temperature parameter using the probability distribution that reaches its maximum of an optimal value instead of directly obtaining a variational problem of a performance index I. This method searches solution spaces in the large, which include solutions that never obtained with the variational method. Therefore, one advantage is a low probability of reaching local solutions. On the other hand, there are some disadvantages. For example, the operation of a cooling schedule, which is a temperature parameter, is difficult. In addition, it may take much time in calculation to converge a solution.

R. Feynman constructed the "path integral" theory in 1948 and showed that it is equivalent to the quantum mechanics. The theory provides a countless number of paths that meet given boundary conditions, and the prob- ability (probability amplitude) of each particle taking its path is expressed in the form of multiple integral [6] . A stochastic behavior is the nature of quantum mechanics. It is therefore impossible to predict a measured physical quantity deterministically under a certain condition. Expected values only can be predicted. Thus, the expected value of a physical quantity is expressed in the form of multiple integral in the path integral method. Furthermore, the multiple integral calculation of an expected value can fit into numerical calculation with the Monte Carlo method.

In this paper, we propose a new numerical calculation method for optimization problems based on the quantum mechanics rather than SA. It means that we show a method of obtaining an approximate solution of an optimal solution as an expected value. An expected value is predicted to be located close to an optimal solution in a continuous probability distribution. This method can thus be applied to various engineering optimal problems.

The introduced probability distribution P here is basically the same as that in SA, and to obtain the minimum of performance function I, the following equation is set:

$$P = e^{-I/h}. \tag{2}$$

In SA, the peak of P is searched with $h \to 0$. On the other hand, we automatically generate a state variable of various time histories at probability P as finite value h. According to the generated frequency, the stochastic average (expected value) of the quantity to be obtained is calculated.

3. FORMULATION OF SPOT

According to the path integral method, the path of motion for a naturally stochastic particle such as photon and electron is obtained from an expected value as a path that keeps the action integral I minimum. We replace the action integral I with a performance index to recognize the motion of a general object as a stochastic process. Thus, we formulate the path integral method for a naturally stochastic particle in physics as an optimal method of obtaining a function that keeps the performance function I minimum. Note that since this performance function is stochastic, an optimal solution is obtained as an expected value, which results in an approximate solution instead of an accurate solution.

According to the path integral method, the action integral $I[x(t)]$ described in the above section is replaced with a performance function to define probability distribution P as shown in Equation (3). Here, $x(t)$ does not always show a path, but shows a general variable to be optimized. Also, A is the normalization constant.

$$P = \frac{1}{A}\exp(-I/h) \quad (3)$$

where h is the Plank constant in quantum mechanics. A stochastic mechanics system matches a classical mechanics system as $h \to 0$. However, in this situation, h is an arbitrary parameter that gives a fluctuation of solution. It means that h is a parameter to define a distribution of P. The larger h is, the wider the distribution is. Also, A is the normalization constant to make the summation of the probability 1. To define the probability distribution as shown in Equation (3), the expected value is obtained from the following:

$$\langle x \rangle = \int_{-\infty}^{\infty} [x(t) \cdot P] Dx(t) \quad (4)$$

where $Dx(t) = \mathrm{d}x_0 \mathrm{d}x_1 \cdots \mathrm{d}x_N$ shows the summation for the paths.

Here, the summation for the paths is the integral for all the countless paths (multiple integral), which for example connect each point x_i with lines at each time point t_i $(0 \le i \le N)$ divided by N in one-dimensional motion. Figure 1 shows some paths that are generated in a probability distribution of Equation (3) when a function defined in Equation (1) is a performance index in one-dimensional upcast motion.

The horizontal axis and vertical axis show time and vertical positions, respectively. The better performance values the paths have, in other words, the closer to the accurate solution (red line) the values are, the higher the density of the paths is. Also, there are some paths that never exist with a deterministic method. The optimal solution is the expected value of all the paths. In this case, that solution is an approximate solution of an accurate solution. The expected solution that is thus obtained shows a path of the highest probability for move of a stochastic behavioral particle, such as an electron, between two points. When those paths are replaced with arbitrary functions, approximate solutions that make the performance function values minimum are obtained.

This idea can be extended to obtain an expected value of a m-dimensional functional that consists of $m \times N$ functions of time in the following way. First, independent variable t is divided by N, and $x_k^{(i)}$ values at time t_i $(0 \le i \le N)$ are randomly chosen. Next, a path that connects these points is set to the time history of variable x_k (Figure 2).

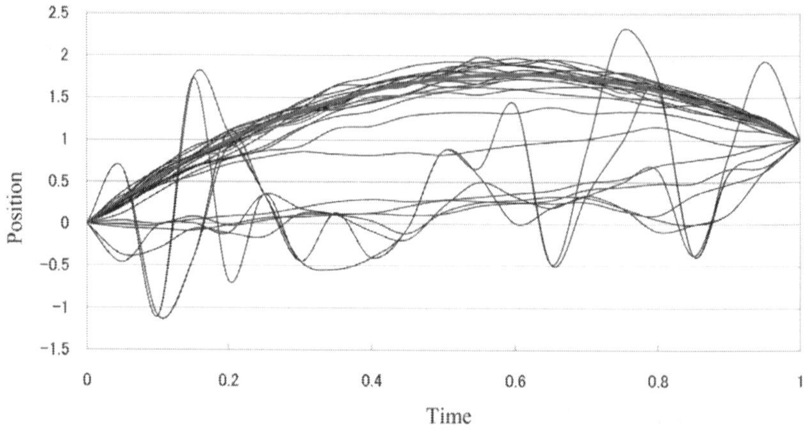

Figure 1. Paths generated by a probability distribution.

Note that the subscript k shows the k-th variable $(1 \le k \le m)$. From Equation (4), the expected value $\langle x_k^{(i)} \rangle$ of x_k at t_i $(0 \le i \le N)$ is expressed as follows:

$$x_k^{(i)} = \int_{-\infty}^{\infty} \left[x_k^{(i)} \cdot P\right] Dx^{(0)}(t)\, Dx^{(1)}(t) \cdots Dx^{(N)}(t),$$
$$Dx^{(i)} = \mathrm{d}x_1^{(i)} \cdot \mathrm{d}x_2^{(i)} \cdots \mathrm{d}x_m^{(i)}. \tag{5}$$

The integral range is from $-\infty$ to ∞ in Equation (5) according to the form of the path integral. The range in actual calculation is between the upper and lower limits of each variable. Also, Equation (5) is formulated to set the values according to the probability distribution for variable $x_k^{(i)}$

at time 0 and N. However, a variable may be fixed in some problems. The end points should be set in accordance with the boundary conditions.

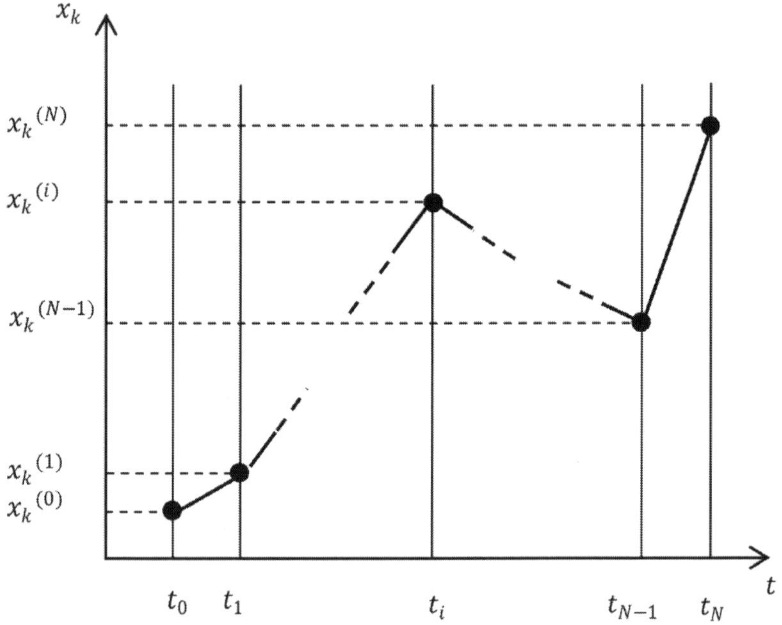

Figure 2. Time history of a variable.

Note that since Equation (5) shows the multiple integral, the Monte Carlo method can be applied to actual numerical calculations.

4. CALCULATION ALGORITHM FOR SPOT

First, a performance index is determined, and if necessary, variables are discretized for numerical calculation. Next, the values of all variables are randomly chosen to generate solutions. Generation probabilities of variables should agree with the probability distribution of Equation (3). Thus, a solution with a fine performance value is generated at a high frequency, and a solution with a poor performance index value is generated at a low frequency. Also, the algorithm is based on the Markov process so that future solutions will have no dependency on the former solutions. Lastly, expected values are calculated using all the solutions generated with a probability distribution to obtain an approximate solution.

The procedure of calculation is shown in Figure 3.

1) The values of all the variables are generated randomly to obtain initial solutions.

2) Time point t_i is chosen according to a random number, and function value x_i at t_i is randomly rewritten to generate a different solution.

3) The performance index value of the newly generated solution is calculated.

4) The adoption or rejection of this solution is decided to generate solutions depending on the probability distribution of Equation (3).

The procedure from Steps 2) to 4) is a pure explanation of the Metropolis method [7] to merely conduct the multiple integral. This method has an advantage of no necessity to obtain the normalization constant: A in Equation (3). A prime characteristic of this method is to calculate the expected values in Step 5).

5) The expected value is calculated.

6) If the final condition is met, the calculation ends. If not, it returns to the generation of solutions.

The final condition can be a certain range of variation of an expected value. However, in this paper, the final condition is the number of iterations, K_{max}. This is because the generation of solutions during the calculation is a Markov process, and therefore, the still more iterations of calculation can result in fine performance index values if the convergence of expected values is not enough.

Equation (3) for probability distribution in this paper is the same as that in SA. SA is a deterministic approach to search only optimal solution at $h \to 0$. On the other hand, h is finite and an optimal solution is obtained as an expected value (stochastic average value) in this paper. It means that the two approaches are based on fundamentally different concepts. In the concrete, SA starts a search with an initial solution for candidate solutions according to the probability distribution. It adopts a solution of the best performance function value as an optimal solution. On the other hand, our method determines an approximate solution to an optimal solution as an expected value with all the solutions generated according to the probability distribution. In this method, there is no process of the so-called cooling schedule which is in SA.

5. CALCULATION EXAMPLE

This section describes examples of numerical calculation with SPOT to demonstrate the validity. First, SPOT is applied to a problem where the minimum value of a multi-peaked function is obtained without staying at other minimal values and the minimum value is always only one with no dependency on the initial value. Next, we apply SPOT to a problem of obtaining the minimum value of a functional to show the validity in a physical optimization problem. Lastly, we apply SPOT to an engineering optimal problem on the glide of a hang glider to show the high applicability of an approximate solution by SPOT.

We already applied SPOT to some engineering problems [8] [9] .

5.1. MINIMIZATION PROBLEM OF BIVARIATE FUNCTION

First, we apply SPOT to a problem of obtaining the minimum value of a bivariate function expressed in Equation (6). Here, performance function I is the function value $f(x,y)$ as follows:

$$f(x,y)=\{\cos(2\pi x)+\cos(2.5\pi x)-2.1\}\times\{2.1-\cos(3\pi y)-\cos(3.5\pi y)\},\quad 0.0\le x\le 3.0,\ 0.0\le y\le 1.5 \tag{6}$$

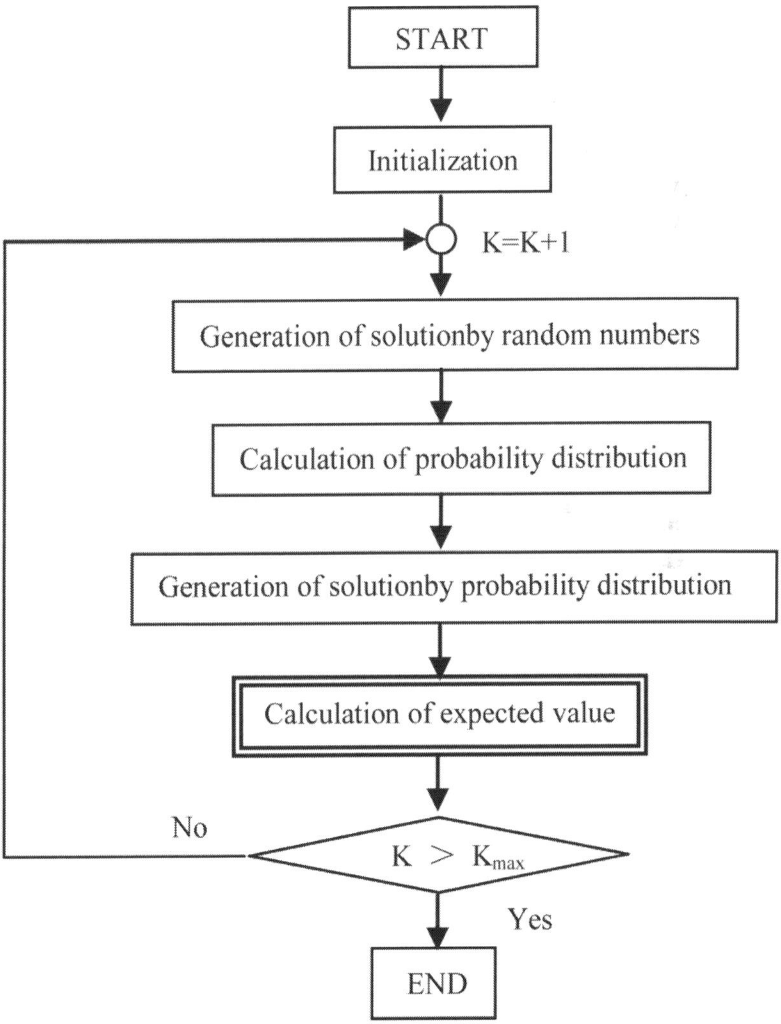

Figure 3. Flow chart of calculation.

Figure 4(a) shows the solution space of Equation (6). The set of $f(x,y)$ that minimizes this function is (0.439, 0.306). This function has multiple minimal values and thus is a multi-peaked function. The contours are shown in the bottom plane, and Figure 4(b) shows the top view.

This example is a problem for obtaining the minimum value of a bivariate function. Therefore, the variables are not discretized, but the values of two variables are generated according to the probability distribution (3) defined by the performance values to obtain an expected value. The closed dots in Figure 4(a) show multiple initial values chosen to check the dependency on the initial values. Also, the cross marks in Figure 4(b) show the minimum values obtained with this method.

If calculation starts with any initial value marked in the closed dot, the resultant values do not stay at any of the many minimal values in the solution space, but all the values concentrate near the minimum value. Since every solution generated in each calculation is obtained stochastically with this method, the calculation results are not influenced by the initial value in principle. However in reality, the number of calculations is finite. Therefore, the expected value as the minimum value may have a variation. The minimum values in Figure 4(b) have a little variation though it may be invisible. This variation is caused by the limited number of calculations.

Note that it took about 18 seconds to perform one million times of calculations on a 1.4 GHz Pentium 4 machine.

5.2. Line of Swiftest Descent Problem

Next, we applied SPOT to a problem on brachistochrone as one of the classical optimization problems to obtain a function. This problem is demonstrated to obtain a mass point path that shows the brachistochrone between a fixed start point and an end point at different altitudes under the uniform gravitation field. Here, performance function I is time t until the mass point reaches the end point as expressed in the following equation:

$$t = \frac{1}{\sqrt{2g}} \int_0^{y_f} \frac{\sqrt{1+\left(dx/dy\right)^2}}{\sqrt{y}} dy \quad (7)$$

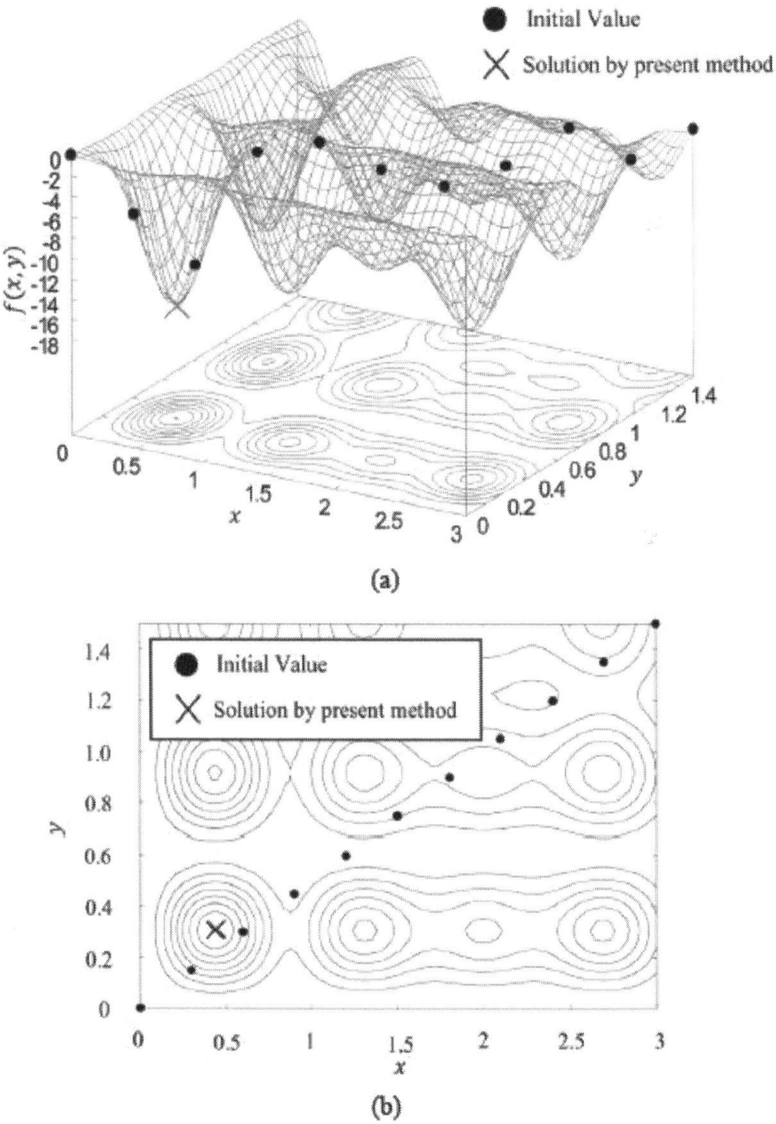

Figure 4. (a) Function space; (b) Contour lines of solution space.

The boundary conditions are as follows:

Initial condition: $y(0)=x(0)=0$, final condition: $y(t_f)=1.0$, $x(t_f)=\pi/2$.

In this problem, we discretized the vertical axis y at an even division, and generate the values of the horizontal axis x corresponding to y according to the probability distribution of Equation (3) defined by the performance index of Equation (7) to obtain an expected value.

Figure 5 shows the paths obtained with SPOT in the closed dots and the analytic solution in the solid line. It took about 438 seconds on a Pentium 4 machine to perform 20 million times of iterations.

SPOT is to obtain an approximate solution of an optimal solution. However, the approximate solutions agree well with the analytic solutions. Thus, this method can be applied to general optimization problems where performance functions are expressed in functional.

5.3. Optimal Glide Problem of Hang Glider

Lastly, we apply SPOT to a problem of a hang glider that starts the fall and glide at an altitude of 12 m toward the ground to obtain its flight operation (time history of angle of attack) for the maximum down range (as shown in Figure 6).

This problem was firstly solved by Suzuki et al. [10] and their results were used to check the validity of the results of SPOT.

This problem is based on the former problem of [10] and the specifications of the hang glider used here are based on those of the optimal glider obtained in [10] . The shape of the hang glider is shown in Figure 7.

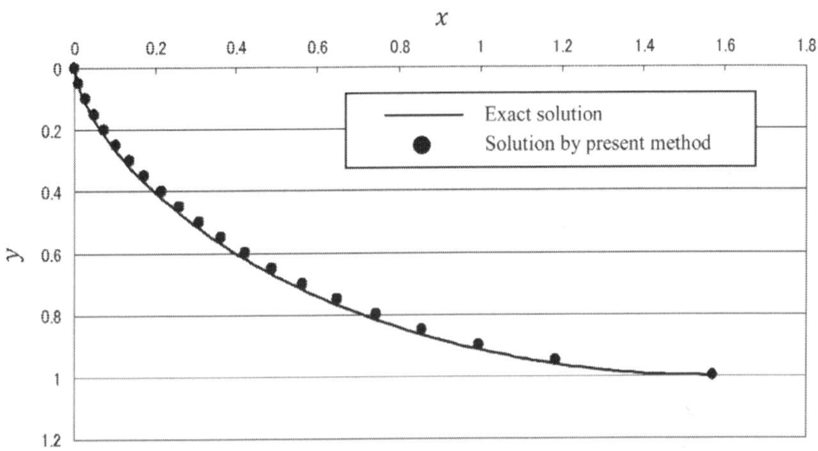

Figure 5. Line of swiftest descent.

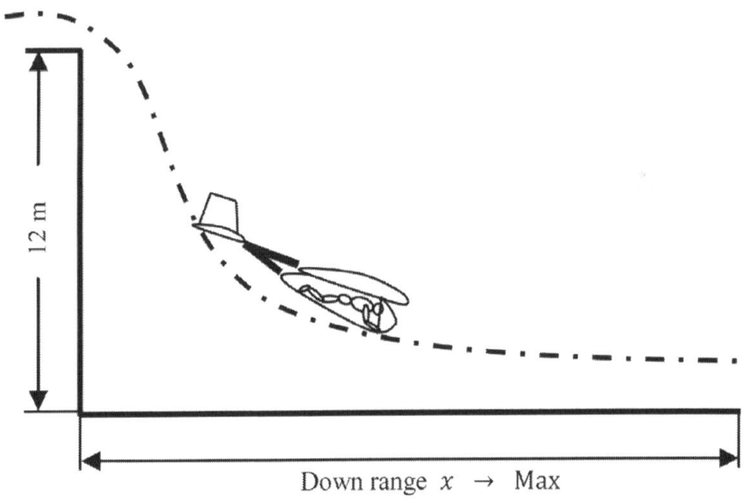

Figure 6. Flight trajectory optimization.

Note that the performance index is the reciprocal number of the flight distance to set this problem to a minimal value problem. Also, the maximum value of the load factor is set to the restriction condition. If this condition exceeds the limitation during the flight, a penalty is added to the performance index value.

The equation of motion is as follows:

$$\dot{V} = -(1/2m)\rho V^2 S_1 C_D^* - g\sin\gamma, \tag{8}$$

$$\dot{\gamma} = (1/2m)\rho V S_1 C_L^* - (g/V)\cos\gamma, \tag{9}$$

$$\dot{x} = V\cos\gamma \tag{10}$$

$$\dot{z} = V\sin\gamma \tag{11}$$

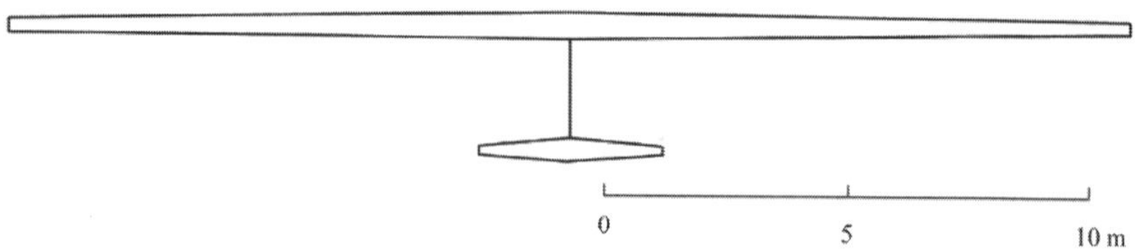

Figure 7. Top view of the hang glider.

The main conditions are shown as follows:

Performance index:

$$I = 1/x_f + \sum_{n_i > n_{\max}} n_i/x_f \tag{12}$$

Initial condition: $z_0 = 12.0$ m, $V_0 = 5.0$ m/s, $\gamma_0 = -3.5$ deg.

Final condition: $z_f = 2.0$ m.

Restriction condition: $n_{\max} = 3.0$.

Specifications of hang glider: $S_1 = 14.6$ m², $A_1 = 36.0$, $b_1 = 22.9$ m, $\lambda_1 = 0.5$, $S_2 = 2.3$ m², $A_2 = 6.0$, $\lambda_2 = 0.5$, $m_0 = 35.0$ kg, $m_p = 60.0$ kg.

The performance index value is the reciprocal number of the down range at the point where the hung glider reaches at an altitude of 2 m. First, the control variable, which is the angle of attack in this case, is discretized on the axis of time. Next, values of angle of attack at each time are specified with random numbers. The equations of motion are numerically calculated according to the time history of angle of attack to obtain the flight path. Then, the performance index value is determined as a reciprocal number of the down range. Therefore, the performance index values are obtained when the series of calculations ends with the final conditions being met.

Figure 8(a) and Figure 8(b) show the relation between the down range and the altitude, and that between the down range and the angle of attack, respectively.

This result may be an approximate solution close to the former optimization [9] . Note that it took about 15 hours on a Pentium 4 (1.4 GHz) machine to obtain the calculation result in this example that set the number of iterations to 3 millions.

6. DISCUSSION

We propose a new optimization method with a stochastic process called SPOT, and then demonstrate numerical calculations.

Generally, when proposing a new method, comparison of performance between the conventional method and a new method should be performed. However, since SPOT is the only method of obtaining the approximate value of the optimal solution, there is no sense in comparing them. Also fair comparison of computation time could not be performed.

As a result, SPOT has the following characteristics:

1) Advantages of Stochastic Process

SPOT results in an expected value from all the solutions generated by calculations. One of the advantages is to avoid influences from the initial value (a start point of the calculation) to the final expected value. The reason is that the generation of solutions during the calculation is a simple stochastic process, and the initial value is only the first-generated solution. As shown in Section 6.1, the calculation results all do not stay at local minimal values even though the calculation starts with any initial value, but the resultant values concentrate near the minimum value. Another advantage is to update an expected value from optimal solutions separately obtained under the same condition. This enables a parallel calculation and the recycling of the former calculation results. This characteristic is unique to SPOT and advantageous to other numerical calculation methods. For example, the calculation of a bivariate function, as shown in Section 6.1, gives the results starting with the multiple initial points as different minimum values. (The cross marks actually show almost the same point.) On the other hand, it is possible to obtain an expected value with all these expected values.

Also, SPOT seems to use asymptotic calculation in the algorithm. However, the former solutions are not used to generate the next solution. This means that the series of calculation is independent and thus the calculation time can easily be estimated with the number of calculations. Therefore, it is easy to estimate the calculation cost.

2) Independency on Experience

Another characteristic is that SPOT has only fluctuation parameter, h, set by operators. There may be a case of extremely low convergence of a solution because of a large fluctuation during calculation with an conventional optimization method using an inclination of the solution curved surface if the features of the solution space are not well known. On the other hand, in principle, the scale of the fluctuation parameter in our method has almost no influence on the process of obtaining an expected value. Although variations of convergence times of calculations due to the scale of the parameter are observed in actual calculations with the limited number of calculations, any large influence on the obtained expected values has not been known. Therefore, this method can be applied to unknown problems, including the surroundings, on solution spaces.

Also, the smaller the fluctuation parameter during calculation, the shorter the calculation time is, though this characteristic is not mentioned in this paper.

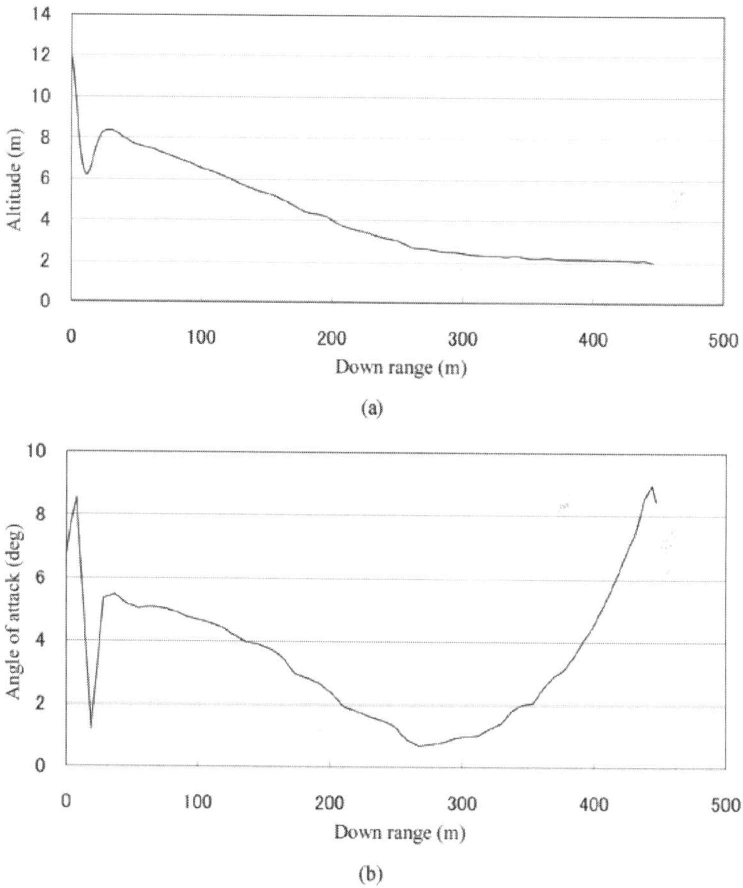

Figure 8. (a) Altitude vs. down range; (b) Angle of attack vs. down range.

3) Restriction Conditions

In the conventional calculation approaches that start with initial solutions to search better performance index values, it may be difficult to generate an initial solution itself and thus to start the numerical calculation if the solution space is narrowed with restriction conditions. However, since the generation of solutions during calculation in our method is a stochastic process, it is possible to perform the calculation regardless of the presence of solutions.

Also, if there are restriction conditions on parameters, which are not the fluctuation parameter: h in (b) but are the load factor: n and the angle of attack: α etc. in Section 6.3 for example, it is necessary to keep parameters within the restriction conditions with general methods. However, with our method, such a device is not necessary. Furthermore, the restriction conditions are rather advantageous to numerical calculations. For example, if there are restrictions on parameter values and the range is specified, solutions can be generated only with values within the restrictions in the process where the series of solutions are generated. This is largely advantageous to perform numerical calculations. The harder the restriction conditions are and the narrower the range of the resultant parameter values is, the more densely the solutions are generated within the restriction conditions and the more easily the calculations are.

Next, in case of restriction conditions on state variables, when a solution generated during calculation is trapped on the restriction, the solution may be abandoned. However, expected values may approach but do not reach the restriction surfaces using only parameter values within constraint ranges and abandoning solutions that break the restrictions. Parameters and solutions that break the restrictions can be applied to a calculation using a penalty method to approach the expected value toward the restriction surface more closely. However, if results with this method are used as initial solutions of deterministic approaches, there is no necessity to make a device.

4) Scope of SPOT

The characteristics of SPOT are to obtain expected values of solution spaces with multiple integral. Thus, there is no problem if the solution spaces have restrictions as mentioned above. However, SPOT is not appropriate to be used for problems that have no restriction on the infinite solution spaces. Also, if a problem has a large solution space and a sharp peak on the performance index value, this method can be applied in principle. However, it may be difficult to obtain an approximate solution with high accuracy in the finite number of calculations. However, it is rare to have infinite range of a variable in an engineering problem, and the upper and lower limits can generally be set. Also, the fluctuation of a performance index value to that of a variable is expected to be smooth. Furthermore, the accuracy required for a variable is finite. Therefore, it is possible to obtain an approximate value for practical use with the finite number of calculations.

7. CONCLUSIONS

We propose a new optimization method based on the path integral called SPOT, and thus perform a numerical calculation on an engineering optimization problem. As a result, we obtain an approximate value of an optimal solution with high accuracy. In addition, this stochastic method has advantages on numerical calculation to other methods.

Hiroaki Yoshida, Katsuhito Yamaguchi, Yoshio Ishikawa Also, SPOT has more effective features of optimization than the conventional methods on optimization problems of an unknown system or narrow solution space that leads to difficulty finding a feasible solution.

REFERENCES

1. Kirkpatrick, S., Gelatt Jr., C.D. and Vecchi, M.P. (1983) Optimization by Simulated Annealing. Science, 220, 671-680.

2. Terasaki, M., Kondo, M., Yoshida, H., Yamaguchi, K. and Ishikawa, Y. (2004) Integrated Optimization of Airplane Design and Flight Trajectory by New Optimization Method Using a Stochastic Process. In: CONPUTATIONAL MECHANICS WCCM VI in Conjunction with APCOM'04, Tsinghua University Press & Springer-Verlag, Beijing, 328.

3. Yoshida, H., Konno, T., Yamaguchi, K. and Ishikawa, Y. (2005) New Optimization Method Using Stochastic Process Based on Path Integral. Journal of the Japan Society for Computational Methods in Engineering, 5, 145-150. (in Japanese)

4. Yoshida, H., Yamaguchi, K. and Ishikawa, Y. (2006) Application of a New Optimization Method for System Design. Joint 3rd International Conference on Soft Computing and Intelligent Systems (SCIS) and 7th International Symposium on Advanced Intelligent Systems (ISIS), Tokyo, 20-24 September 2006, 1542-1547.

5. Yoshida, H., Yamaguchi, K. and Ishikawa, Y. (2007) Design Tool Using a New Optimization Method Based on a Stochastic Process. Transactions of the Japan Society for Aeronautical and Space Sciences, 49, 220-230.

6. Feynman, R.P. and Hibbs, A.R. (1965) Quantum Mechanics and Path Integrals. McGraw-Hill, New York.

7. Metropolis, N., Rosenbluth, A.W., Rosenbluth, M.N., Teller, A.H. and Teller, E. (1953) Equation of State Calculations by Fast Computing Machines. Journal of Chemical Physics, 21, 1087-1092.

8. Nakane, M., Kobayashi, D., Yoshida, H. and Ishikawa, Y. (2009) Feasibility Study on Single Stage to Orbit Space Plane with RBCC Engine. 16th AIAA/DLR/DGLR International Space Planes and hypersonic Systems and Technologies Conference, Bremen, AIAA 2009-7331.

9. Ishimori, Y., Nakane, M., Ishikawa, Y., Yoshida, H. and Yamaguchi, K. (2011) Feasibility Study for Spaceplane's Concepts in Ascent Phase Using the Optimization Method. Journal of the Japan Society for Aeronautical and Space Sciences, 59, 291-297. (in Japanese)

10. Suzuki, S. and Kawamura, N. (1996) Simultaneous Optimization of Sailplane Design and Its Flight Trajectory. Journal of Aircraft, 33, 567-571.

CHAPTER 2

Multiyear Discrete Stochastic Programming with a Fuzzy Semi-Markov Process

C. S. Kim[1], Richard M. Adams[2], Dannele E. Peck[3]

[1] Economic Research Service, U.S. Department of Agriculture, Washington, DC, USA

[2] Department of Applied Economics, Oregon State University, Corvallis, OR, USA

[3] Department of Agricultural and Applied Economics, University of Wyoming, Laramie, WY, USA

ABSTRACT

Drought conditions at a given location evolve randomly through time and are typically characterized by severity and duration. Researchers interested in modeling the economic effects of drought on agriculture or other water users often capture the stochastic nature of drought and its conditions via multiyear, stochastic economic models. Three major sources of uncertainty in application of a multiyear discrete stochastic model to evaluate user preparedness and response to drought are: (1) the assumption of independence of yearly weather conditions, (2) linguistic vagueness in the definition of drought itself, and (3) the duration of drought. One means of addressing these uncertainties is to re-cast drought as a stochastic, multiyear process using a "fuzzy" semi-Markov process. In this paper, we review "crisp" versus "fuzzy" representations of drought and show how fuzzy semi-Markov processes can aid researchers in developing more robust multiyear, discrete stochastic models.

KEYWORDS

Drought, Discrete Stochastic Economic Modeling, Fuzzy Logic, Fuzzy Markov Process, Fuzzy Semi-Markov Process

1. INTRODUCTION

The impacts of drought on water supplies and those dependent on such supplies have long been an important topic for economists and policy makers. Climate projections for key hydrologic inputs—seasonal precipitation, snowpack storage, evaporative loss, and frequency and severity of drought—are used to anticipate future stresses on water quantity and variability. Water managers use these projections to allocate seasonal water sup plies across various classes of users. For example, public irrigation districts in the western U.S. typically issue water allocation declarations in March, based on hydrologic forecasts, to help irrigators plan for the coming irrigation season. In periods of sharply reduced water supplies, such as recent multiyear droughts in California (**Figure 1**), such allocation decisions must reflect a range of competing demands, including in stream flow requirements to protect associated species and ecosystems.

Discrete stochastic programming (DSP) has been widely used as an optimization method to assess the benefits and costs of alternative water allocation decisions during varying degrees of drought [1]-[3]. DSP is a useful technique based on traditional probability theory. To illustrate water supply uncertainty in a discrete stochastic programming framework, first consider the following single-year linear programming model:

$$\text{Max } Z = c'x, \tag{1.1}$$

subject to:

$$Ax \le b \tag{1.2}$$

where $x = (xj)$ is a vector of control variables with $x \ge 0$, $c = (cj)$ is an $(n \times 1)$ column vector of constants, $A = (aij)$

where $\sum_{k=1}^{K} P_k = 1$.

2. LINGUISTIC VAGUENESS AND FUZZY LOGIC

To evaluate the impacts of drought conditions and associated preparedness and response plans, a clear definition of drought must be provided. One definition of drought is the case in which irrigation water supplied is less than irrigation water demanded, due to inadequate rainfall, snow pack, or other weather conditions. As the difference between irrigation water demanded and supplied increases, severity of drought intensifies along a continuous gradient. While the characterization of drought varies across studies, the following definition of drought pro- vided by Yevjevich [14] has been widely used [6] [15] [16] :

$$y_i = \begin{cases} 0 & \text{if } S_i < S_0 \quad (i = 1, 2, 3, \cdots) \\ 1 & \text{if } S_i \ge S_0 \end{cases}, \tag{2.1}$$

where S_0 is a constant water allotment threshold (such as 40 acre-inches Peck and Adams [3] used for water allotment under normal weather condition) and S_i is the ith severity state. A Bernoulli variable y_i plays a significant role in estimation of the holding time (i.e., duration) probability mass function of drought in later section of semi-Markov chains.

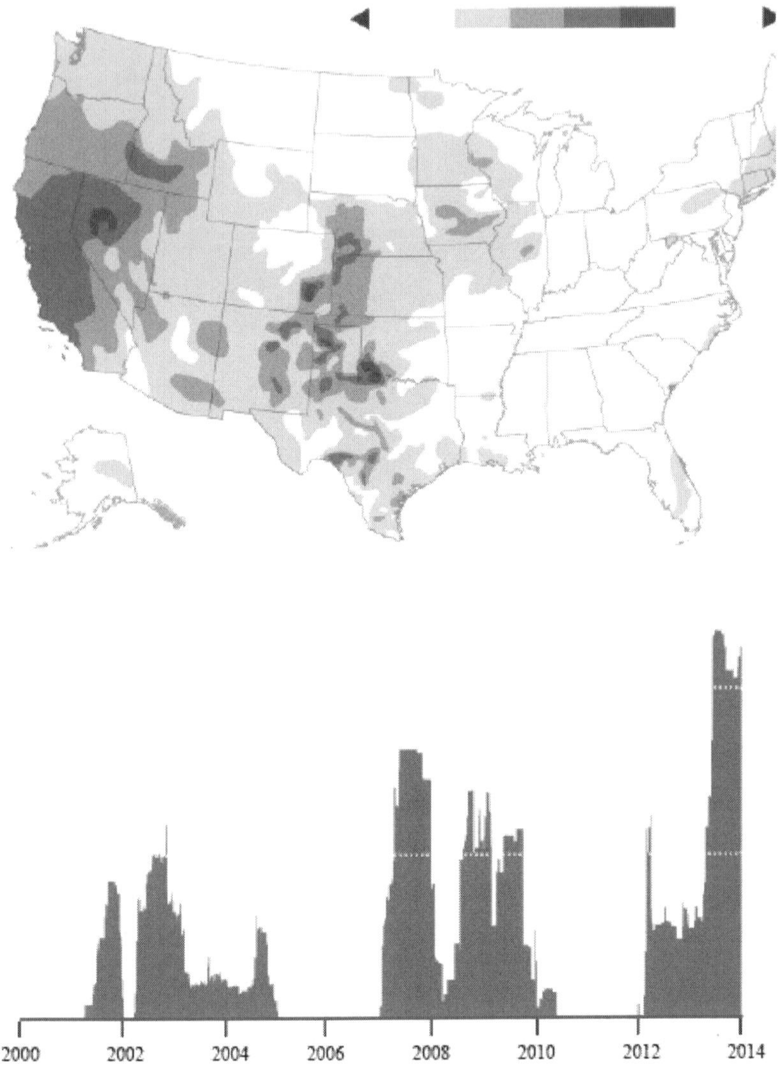

Figure 1. Severity and duration of drought in California (Source: The National Drought Miti- gation Center—USDA).

When economic models represent drought in binary terms (i.e., a water allocation either qualifies or does not qualify as drought), this overly-simplified or deceivingly-crisp (as opposed to fuzzy) measurement of drought severity can cause inefficient resource allocation. Unlike this crisp set (in which an element is either a member of the set or not), fuzzy sets allow elements to be included through a degree of membership, as expressed by a membership function, thus relaxing the binary state assumption [6] [17] . Introduced by Zadeh [18] , fuzzy logic and fuzzy set theory have since been widely adopted to deal with linguistic vagueness in various mathematical optimization models, including fuzzy dynamic programming [19] [20] , and optimal fuzzy control [21] .

$$f_s(w_i) = \begin{cases} 1 & \text{if } w_i \geq w_0 \\ 1 - \dfrac{(w_0 - w_i)}{\alpha} & \text{if } w_0 - \alpha \leq w_i < w_0 \text{.} \\ 0 & \text{otherwise} \end{cases} \tag{2.2}$$

3. FUZZY MARKOV AND SEMI-MARKOV PROCESSES

Definition 2. Let $W = \{w_1, w_2, \cdots, w_n\}$ be a finite state-space of possible water allotments from an irrigation water district. Let $(S_n)_{n \in N}$ be random variables taking values in W. Now, let a fuzzy subset of W be defined as

a set of ordered pairs $(S_i, f_S^m(S_i))$, where $f_S^m(S_i)$ is the grade of membership of S_i in W in the mth transition.

$$f_S^{(m+1)}(S_j) = \max_{i \in S}\left\{\min\left(f_S^m(S_i), P_{ij}\right)\right\} \quad j \in S \text{,} \tag{3.2}$$

where the fuzzy-state grade of membership, $f_S^m(S_i)$, can be calculated as follows:

$$f_S^m(S_i) = \max_{l \in S}\left[\min\left(f_S^0(S_l), P_{li}^m\right)\right] \quad i \in S \text{,} \tag{3.3}$$

and the powers of the fuzzy transition matrix, P_{li}^m, are defined by:

$$P_{li}^m = \max_{k \in S}\left[\min\left(P_{lk}, P_{ki}^{m-1}\right)\right] \text{,} \tag{3.4}$$

where $P_{ii}^1 = P_{ii}$ and P_{ii}^0 is an identity matrix at the initial time period.

A few empirical studies have been conducted to compare efficiencies between the use of classic Markov chains based on conventional crisp set theory and fuzzy Markov chains in the context of stochastic programming models. Mousavi et al. [20] compared a conventional stochastic dynamic programming model (which employed classic Markov chains) and a fuzzy-state stochastic dynamic programming model (which employed fuzzy Markov chains) in the operation of a multipurpose reservoir in Iran. Results from their study show that fuzzy-state stochastic dynamic programming outperforms conventional stochastic dynamic programming in achieving the flood control objective and in overall performance of the reservoir. Chandramouli and Nanduri [27] also compared a conventional stochastic dynamic model and a fuzzy-state stochastic dynamic model in the operation of a multipurpose reservoir in India. They found that the fuzzy-state stochastic dynamic programming model out performed the conventional stochastic dynamic programming model.

3.1. Fuzzy Semi-Markov Process

3.1.1. Homogeneous Fuzzy Semi-Markov Process

Where S^n represents the state at the nth transition, T_n is the time of the nth transition and "t" is arrival time. Equation (3.5) expresses the probability of transitioning to state j at arrival time t, given the system has been in state i after n transitions. Unfortunately no solution has been reported for Equation (3.5) in the literature; therefore, an alternative approach for solving Equation (3.5) has been employed by Cancelliere and Salas [15] and Mirakbari and Ganji [6] .

$$h_{ij}\left(d_{ij}\right) = r_{ij}\left(1 - r_{ij}\right)^{d_{ij}-1} \quad \text{for } d_{ij} = 1, 2, 3, \cdots, \tag{3.9}$$

where r_{ij} represents the probability of the duration d_{ij} in the ith drought severity, before moving to the jth drought severity. This is estimated by $r_{ij} = \frac{1}{\bar{d}_{ij}}$, where $\bar{d}_{ij}$ is the mean holding-time (i.e., duration) in the ith

drought before moving to the jth drought, which can be easily observed from historical data.

Mirakbari and Ganji [6] compared performances between the classic semi-Markov chains and the fuzzy semi- Markov chains by using profust reliability theory to a rangeland system in India.6 Their results indicate that the reliability of rangeland system decisions increased by 22 percent when the fuzzy semi-Markov process is used over the classic semi-Markov process.

3.1.2. Non-Homogeneous Fuzzy Semi-Markov Process

Where q_{ij} is the transition probability of the embedded fuzzy Markov chain in the process.

$$\begin{aligned} P_i(s, d_i) &= P\left[T_{n+1} - T_n \leq d_i \mid S^n = w_{i,n}, T_n = s\right] \\ &= \max_{j \neq i}\left\{\min\left(P_{ij}(s), h_{ij}(s, d_{ij})\right)\right\}. \end{aligned} \tag{3.17}$$

4. NUMERICAL EXAMPLE

We now present a numerical example of multiyear water supply forecasts under the assumption that a pair of two random variables $\{S_n, T_n\}$ follows a homogeneous fuzzy semi-Markov process (i.e., Equation (3.5) through (3.12)).7 The multiyear forecast associated with the severity and duration of droughts can be incorporated into a multiyear discrete stochastic programming model as shown in Equation (3.12).

Table 1. Transition probabilities and holding-time mass functions.

Equation	*Transition probability*	*Observed mean duration*	*Holding-time mass function*
$(A.1)$	$p_{11} = 0.4$	$\bar{d}_{11} = 3$	$h_{11}(d_{11}) = \left(\frac{1}{3}\right)\left(\frac{2}{3}\right)^{d_{11}-1}$
$(A.2)$	$p_{12} = 0.4$	$\bar{d}_{12} = 4$	$h_{12}(d_{12}) = \left(\frac{1}{4}\right)\left(\frac{3}{4}\right)^{d_{12}-1}$
$(A.3)$	$p_{13} = 0.2$	$\bar{d}_{13} = 5$	$h_{13}(d_{13}) = \left(\frac{1}{5}\right)\left(\frac{4}{5}\right)^{d_{13}-1}$
$(B.1)$	$p_{21} = 0.3$	$\bar{d}_{21} = 3$	$h_{21}(d_{21}) = \left(\frac{1}{3}\right)\left(\frac{2}{3}\right)^{d_{21}-1}$
$(B.2)$	$p_{22} = 0.5$	$\bar{d}_{22} = 3$	$h_{22}(d_{22}) = \left(\frac{1}{3}\right)\left(\frac{2}{3}\right)^{d_{22}-1}$
$(B.3)$	$p_{23} = 0.2$	$\bar{d}_{23} = 4$	$h_{23}(d_{23}) = \left(\frac{1}{4}\right)\left(\frac{3}{4}\right)^{d_{23}-1}$
$(C.1)$	$p_{31} = 0.4$	$\bar{d}_{31} = 4$	$h_{31}(d_{31}) = \left(\frac{1}{4}\right)\left(\frac{3}{4}\right)^{d_{31}-1}$
$(C.2)$	$p_{32} = 0.3$	$\bar{d}_{32} = 5$	$h_{32}(d_{32}) = \left(\frac{1}{5}\right)\left(\frac{4}{5}\right)^{d_{32}-1}$
$(C.3)$	$p_{33} = 0.3$	$\bar{d}_{33} = 6$	$h_{33}(d_{33}) = \left(\frac{1}{6}\right)\left(\frac{5}{6}\right)^{d_{33}-1}$

$$H\left(d_{ij}\right)=\begin{matrix} & N & MM & SEE \\ N & 1-\left(\frac{2}{3}\right)^{d_{11}} & 1-\left(\frac{3}{4}\right)^{d_{12}} & 1-\left(\frac{4}{5}\right)^{d_{13}} \\ MM & 1-\left(\frac{2}{3}\right)^{d_{21}} & 1-\left(\frac{2}{3}\right)^{d_{22}} & 1-\left(\frac{6}{7}\right)^{d_{23}} \\ SEE & 1-\left(\frac{3}{4}\right)^{d_{31}} & 1-\left(\frac{4}{5}\right)^{d_{32}} & 1-\left(\frac{5}{6}\right)^{d_{33}} \end{matrix}. \quad (4.3)$$

Comparison

Due to the lagging and long-term effects of drought on vegetation and soil moisture, or on cropping choices due to agronomic constraints (e.g., rotations), the resilience of drought is equal to or longer than drought duration [38] .9 Therefore, economic impacts of individual years of drought may not necessarily be independent [3] . To compare multiyear drought probabilities between a conventional multiyear discrete stochastic program (which typically assumes the previous year's state of nature does not influence the probability of future states of nature) and a multiyear homogeneous fuzzy semi-Markov process, the steady-state probabilities of the fuzzy transition matrix in Equation (4.2) are estimated as follows:

$$\pi=\begin{matrix} N & MM & SEE \\ \left[\frac{1}{3},\right. & \frac{1}{3}, & \left.\frac{1}{3}\right] \end{matrix}, \quad (4.4)$$

where π is a stationary distribution.1[0] The probabilities in Equation (4.4) represent P_k (k = 1, 2, 3) in Equation (1.3). In a conventional discrete stochastic programming model, the probability of severe drought (SEE) in the first year, as well as in the second year, or any other year, is $\frac{1}{3}$, such that the probability of two consecutive years of severe drought is $\left(\frac{1}{3}\right)^2=0.1111$, while the probability of three consecutive years of severe drought is $\left(\frac{1}{3}\right)^3=0.0370$.

Table 2. Results of a numerical example.

Weather	Model	Duration		
		d = 1	d = 2	d = 3
N	FsM	$P_1(d = 1) = 0.3333$	$P_1(d = 2) = 0.2222$	$P_1(d = 3) = 0.1481$
	DSPM	0.3333	0.1111	0.0370
MM	FsM	$P_2(d = 1) = 0.3000$	$P_2(d = 2) = 0.2222$	$P_2(d = 3) = 0.1481$
	DSPM	0.3333	0.1111	0.0370
SEE	FsM	$P_3(d = 1) = 0.2500$	$P_3(d = 2) = 0.1875$	$P_3(d = 3) = 0.1406$
	DSPM	0.3333	0.1111	0.0370

N = Normal; MM = Mild/Moderate drought; SEE = Severe/Extreme/Exceptional drought. FsM = Fuzzy semi-Markov; DSPM = Discrete Stochastic Programming Model.

The possibility mass function of the holding-time spent in each weather condition in a fuzzy semi-Markov process is estimated with Equation (3.11) and results are presented in Table 2. In contrast to a conventional discrete stochastic programming model, the possibility of two consecutive years of SEE drought condition is 0.1875, which is 69 percent higher than the probability assumed in a conventional multiyear discrete stochastic programming approach. As the duration of consecutive years of severe drought increases to three years, the possibility of holding time in a fuzzy semi-Markov process is 0.1406, which is 280 percent higher than the probability assumed in a conventional multiyear discrete stochastic programming approach, 0.0370.

Implementation of adequate measures to control or mitigate drought consequences is a major challenge for irrigators and other water users. Our numerical example, while stylized, demonstrates how economists can use fuzzy semi-Markov processes to incorporate uncertainty about both severity of drought (which necessitates fuzzy sets) and duration of a multiyear drought (which necessitates semi-Markov processes) in stochastic modeling by using a fuzzy semi-Markov process. Such model specification may improve representation of the economic effects of drought severity and duration on water users and the efficacy of alternative mitigation actions.

ACKNOWLEDGEMENTS

The views expressed are those of the authors and should not be attributed to USDA or Economic Research Service.

REFERENCES

1. Engelhardt, T. (1983) Water Supply Optimization with Discrete Stochastic Linear Programming. Proceedings of International Crops Research Institute Workshop on the State of the Art and

Management Alternatives for Optimizing the Productivity of SAT Alfisols and Related Soils, ICRISAT Center, 1-3 December 1983, Patancheru, India, 79-87.

2. McCarl, B. and Parandvash, G. (1988) Irrigation Development versus Hydroelectric Generation: Can Interruptible Irrigation Play a Role? Western Journal of Agricultural Economics, 13, 267-276.

3. 3. Peck, D.E. and Adams, R.M. (2010) Farm-Level Impacts of Prolonged Drought: Is a Multiyear Event More than the Sum of Its Parts? The Australian Journal of Agricultural and Resource Economics, 54, 43-60.

4. Cocks, K.D. (1968) Discrete Stochastic Programming. Management Science, 15, 72-79.

5. Watanabe, Y. (1987) Errors in Availability Estimation by 2-State Models of 3-State Systems. Reliability Engineering, 18, 223-235.

6. Mirakbari, M. and Ganji, A. (2010) Reliability Analysis of a Rangeland System: The Application of Profust Theory. Stochastic Environmental Research and Risk Assessment, 24, 399-409.

7. Sahinidis, N. (2004) Optimization under Uncertainty: State-of-the-Art and Opportunities. Computers and Chemical Engineering, 28, 971-983.

8. Fullér, R. (1998) Fuzzy Reasoning and Fuzzy Optimization. Department of Operations, Eötvös Loránd University, Budapest.

9. Inuiguchi, M. and Ramit, J. (2000) Possibilistic Linear Programming: A Brief Review of Fuzzy Mathematical Programming and a Comparison with Stochastic Programming in Portfolio Selection Problem. Fuzzy Sets and System, 111, 3-28.

10. Tanaka, H., Okuda, T. and Asai, K. (1974) Decision Making and Information in Fuzzy Events. Bulletin Series A, 23, 193-202

11. Li, H. and Gong, Z. (2010) Fuzzy Linear Programming with Possibility and Necessity Relation. In: Cao, et al., Eds., Fuzzy

Information and Engineering, Springer-Verlag, Berlin Heidelberg, 305-311.

12. Huang, Y.F., Baetz, B.W., Huang, G.H. and Liu, L. (2002) Violation Analysis for Solid Waste Management Systems: An Interval Fuzzy Programming Approach. Journal of Environmental Management, 65, 431-446.

13. Arkov, V., Kulikov, G.G. and Breikin, T.V. (1999) Fuzzy Markov Modeling in Automatic Control of Complex Dynamic Systems. International Conference on Accelerator and Large Experimental Physics Control Systems, 4-8 October 1999, Trieste, 287-289.

14. Yevjevich, V. (1967) An Objective Approach to Definitions and Investigations to Continental Hydrologic Droughts. Colorado State University, Fort Collins, Colorado.

15. Cancelliere, A. and Salas, J.D. (2004) Drought Length Properties for Periodic-Stochastic Hydrologic Data. Water Resources Research, 40, W02503.

16. Llamas, J. and Siddiqui, M. (1969) Runs of Precipitation Series. Hydrology Paper 33, Colorado State University, Fort Collins.

17. Cai, K.Y., Wen, C.U. and Zhang, M.L. (1993) Fuzzy States as a Basis for a Theory of Fuzzy Reliability. Journal of Microelectronics Reliability, 33, 2253-2263.

18. Zadeh, L.A. (1965) Fuzzy Sets. Information and Control, 8, 338-353.

19. Bellman, R.E. and Zadeh, L.A. (1970) Decision-Making in a Fuzzy Environment. Management Science, 17, 141-164.

20. Mousavi, S., Karamouz, K. and Menhadj, M.B. (2004) Fuzzy-State Stochastic Dynamic Programming for Reservoir Operation. Journal of Water Resources Planning and Management, 130, 460-470.

21. Baten, M.A. and Kamil, A.A. (2012) Optimal Fuzzy Control with Application to Discounted Cost Production Inventory Planning

Problem. Proceedings of the World Congress on Engineering, Vol. III, London, 30 June-2 July 2012.

22. Khedun, C.P. and Singh, V. (2014) Drought Risk Management in Agriculture: Current Trends in Research. Texas A&M University, College Station.

23. Alves, A., Mota, G., Costa, G. and Feitosa, R. (2012) Estimation of Transition Possibilities for Fuzzy Markov Chains Applied to the Analysis of Multitemporal Image Sequences. Proceedings of the 4th GEOBIA, Rio de Janeiro, 7-9 May 2012.

24. Praba, B., Sujatha, R. and Srikrishna, S. (2009) A Study on Homogeneous Fuzzy Semi-Markov Model. Applied Mathematical Science, 3, 2453-2467.

25. Avranchenkov, K. and Sanche, E. (2002) Fuzzy Markov Chains and Decision-Making. Fuzzy Optimization and Decision Making, 1, 143-159.

26. Schweizer, P.J. (1968) Pertubation Theory and Finite Markov Chains. Journal of Applied Probability, 5, 401-413.

27. Chandramouli, S. and Nanduri, U. (2011) Comparison of Stochastic and Fuzzy Dynamic Programming Models for the Operation of a Multipurpose Reservoir. Water and Environment Journal, 25, 547-554.

28. Bardossy, A. and Plate, E.J. (1991) Modeling Daily Rainfall Using a Semi-Markov Representation of Circulation Pattern Occurrence. Journal of Hydrology, 122, 33-47.

29. Masruroh, N. (2012) Introduction to Discrete Time Semi Markov Process.

30. Praba, B., Sujatha, R. and Srikrishna, S. (2009) Fuzzy Reliability Measures of Fuzzy Probabilistic Semi-Markov Model. International Journal of Recent Trends in Engineering, 2, 25-29.

31. Vassiliou, P.C. (2013) Fuzzy Semi-Markov Migration Process in Credit Risk. Fuzzy Sets and Systems, 223, 39-58.

32. De Medici, J.J. and Manca, R. (2012) The Aggregate Claim Amount Discrete Time Semi-Markov Model. Working Paper, Department of Mathematics of Economic Decisions on Finance and Insurance, University of Rome, La Sapienza.

33. Janssen, J., Manca, R. and di Prignano, E.V. (2012) Non Homogeneous Interest Rate Structure in a Semi-Markov Framework. Working Paper. http://www.actuaries.org

34. Gildeh, B.S. and Dadgar, A. (2007) Reducible Fuzzy Markov Chain and Fuzzy Absorption Probability. 1st Joint Congress on Fuzzy and Intelligent Systems, Ferdowsi University of Mashhad, Iran, 29-31August 2007.

35. Li, J. and He, P. (2010) The Application of Fuzzy Markov Chains in the Analysis of Internet Glance Behavior. 3rd IEEE International Conference on Computer Science and Information Technology (ICCSIT), Vol. 7, Chengdu, 9-11 July 2010, 608-611.

36. Sanchez, E. (1976) Resolution of Composite Fuzzy Relation Equation. Information and Control, 30, 38-48.

37. Meenakshi, A.R. and Kaliraja, M. (2010) Regular Interval Valued Fuzzy Matrices. Advances in Fuzzy Mathematics, 5, 7-15.

38. Rossi, G., Benedini, M., Tsakiris, G. and Giakoumakis, S. (1992) On Regional Drought Estimation and Analysis. Water Resources Management, 6, 249-277.

NOTES

[1] Both possibilistic and flexible fuzzy programming can be presented in interval fuzzy programming [12] .

[2] However, we use the terms possibility and probability interchangeably in this paper.

[3] The max-min operations of fuzzy sets are as follows. For fuzzy subsets A and B of a crisp set W $\neq \phi$, the intersection of A and B is defined as: $(A \cap B)(x) = \min\{A(x), B(x)\} = A(x) \wedge B(x)$, $\forall x \in W$. Similarly, the union of A and B is defined as follows: $(A \cup B)(x) = \max\{A(x), B(x)\} = A(x) \vee B(x)$, $\forall x \in W$.

[4] The set of all subsets of W is called the power set of W.

[5] Markov kernel is also called a transition probability function that maps from a measureable space to another space.

[6] Profust reliability theory consists of two parts, the fuzzy part which considers vagueness in rangeland system failure and the probabilistic part which incorporates randomness of rangeland failure [6] [17] .

[7]Multiyear water supply forecast is a forecast in $d_{ij} = 1$, $d_{ij} = 2$, $d_{ij} = 3$, etc. for all drought severities $i = 1, 2, \times\times\times, n$.

[8] See Avrachenkov and Sanchez [25] , Gildeh and Dadgar [34] , Li and He [35] , Mousavi et al. [20] , and Sanchez [36] for fuzzy transition matrix, and Meenakshi and Kaliraja [37] for interval transition matrix.

[9] Resilience is a measure of the recovery time of the system [38] .

[10] The stationary distribution must satisfy $\lim_{k \to \infty} P^k = 1\pi$, where 1 is a column vector with all entries equals to one.

CHAPTER 3

On the Stability of Stochastic Jump Kinetics

Stefan Engblom

Division of Scientific Computing, Department of Information Technology, Uppsala University, Uppsala, Sweden

ABSTRACT

Motivated by the lack of a suitable constructive framework for analyzing popular stochastic models of Systems Biology, we devise conditions for existence and uniqueness of solutions to certain jump stochastic differential equations (SDEs). Working from simple examples we find reasonable and explicit assumptions on the driving coefficients for the SDE representation to make sense. By "reasonable" we mean that stronger assumptions generally do not hold for systems of practical interest. In particular, we argue against the traditional use of global Lipschitz conditions and certain common growth restrictions. By "explicit", finally, we like to highlight the fact that the various constants occurring among our assumptions all can be determined once the model is fixed. We show how basic long time estimates and some limit results for perturbations can be derived in this setting such that these can be contrasted with the corresponding estimates from deterministic dynamics. The main complication is that the natural path-wise representation is generated by a counting measure with an intensity that depends nonlinearly on the state.

KEYWORDS

Nonlinear Stability, Perturbation, Continuous-Time Markov Chain, Jump Process, Uncertainty, Rate Equation

1. INTRODUCTION

The observation that detailed modeling of biochemical processes inside living cells is a close to hopeless task is a strong argument in favor of stochastic models. Such models are often thought to be more accurate than conventional rate-diffusion laws, yet remain more manageable than, say, descriptions formed at the level of individual molecules. Indeed, several studies [1] -[3] have showed that noisy models have the ability to capture relevant phenomena and to explain actual, observed dynamics.

In this work we shall consider some "flow" properties of a stochastic dynamical system in the form of a quite general continuous-time Markov chain. Since the pioneering work of Gillespie [4] [5] , in the Systems Biology context this type of model is traditionally described in terms of a (chemical) master equation (CME). This is the forward Kolmogorov equation of a certain jump stochastic differential equation (jump SDE for brevity), driven by independent point processes with state-dependent intensities. Despite the popularity of the master equation approach, little analysis on a per trajectory-basis of actual models has been attempted.

In the general literature, when discussing existence/uniqueness and various types of perturbation results, different choices of assumptions with different trade-offs have been made. One finds that the treatment often falls into one of two categories taking either a "mathematical" or a "physical" viewpoint. Both of the conditions are highly general but with subsequently less transparent proofs and resulting in more abstract bounds. Or the conditions are formed out of convenience, say, involving global Lipschitz constants, and classical arguments carry through with only minor modifications.

Protter ([6] Chap. V) offers a nice discussion from the mathematical point of view and in ascending order of generality, including the arguably highly unrestrictive assumption of locally Lipschitz continuous coefficients. Other authors ([7] , Chap. 6, [8] , Chap. 3-5) also treat the evolution of general jump-diffusion SDEs in conti- nuous state spaces.

A study of the flow properties of jump SDEs is found in [9] , where the setting is scalar and the state is continuous. In [10] jump stochastic partial differential equations are treated, and existence/uniqueness results as well as ergodic results for the case of a multiplicative noise, are found in [11] [12] . Numerical aspects in a similar setting are discussed in [13] .

In a more applied context, stability is often thought of as implied from physical premises and the solution is tactically assumed to be confined inside some bounded region ([14] , Chap. V). The fundamental issue here is that for open systems in a stochastic setting, there is a non-zero probability of reaching any finite state and global assumptions must be formed with great care. The analysis of open networks under an a priori assumption of boundedness is therefore quite difficult to interpret other than in a qualitative sense. Notable examples in this setting include time discretization strategies [15] [16] , time-parallel simulation techniques [17] , and parameter perturbations [18] .

Evidently, essentially no systems of interest satisfy global Lipschitz assumptions since the fundamental inte- raction almost always takes the form of a quadratic term. Interestingly, for ordinary differential equations, it has been shown [19] that Lipschitz continuous coefficients imply a computationally polynomial-space complete solution; thus providing a kind of explanation for the convenience with this weak feedback assumption. It is also known [20] , that with SDEs, superlinearly growing coefficients may in fact cause the forward Euler method to diverge.

1.1. Agenda

Besides its expository material, the purpose of this paper is to devise simple conditions that imply stability for finite and, in certain cases, infinite times, and that, when applied to systems of practical interest, yield explicit expressions for the associated stability estimates. As a result the framework developed herein applies in a constructive way to any chemical network, of arbitrary size and topology, formed by any combination of the elementary reactions (2.3) to be presented in Section 2. Additionally, it will be clear how to encompass also other types of nonlinear reactions that typically result from adiabatic simplifications.

As an argument in favor of this bottom-up approach one can note that, for evolutionary reasons, biochemical systems tend to operate close to critical points in phase-space where the efficiency is the highest. Clearly, for such dynamical systems, an analysis by analogy might be highly misleading.

We also like to argue that our results are of interest from the modeling point of view. Due to the type of phenomenological arguments often involved, judging the relative effect of the (non-probabilistic) epistemic uncer- tainty is a fundamental issue which has so far not rendered a consistent analysis.

1.2. Outline

The expository material in Section 2 is devoted to formulating the type of processes we are interested in. We state the master equation as well as the corresponding jump SDE and we also look at some simple, yet informative actual examples. Since it is expected that the properties of the stochastic dynamics are somehow similar to those of the deterministic version, we search for a set of minimal assumptions in the latter setting in Section 3. Techniques for finding explicit values of the constants occurring among our assumptions are also devised. The main results of the paper are found in Section 4 where we put our theory together and prove existence and uniqueness, as well as long time estimates and limit results for perturbations. A concluding discussion is found in Section 5.

2. STOCHASTIC JUMP KINETICS

In this section we start with the physicist's traditional viewpoint of pure jump processes and write down the governing master equations. These are evolution equations for the probability densities of continuous-time Markov chains over a discrete state space. Although the application considered here is mesoscopic chemical kinetics, identical or very similar stochastic models are also used in Epidemiology [21] , Genetics [22] and Sociodynamics [23] , to name just a few.

We then proceed with discussing a path-wise representation in terms of a stochastic jump differential equation. The reason the sample path representation is interesting is the possibility to reason about flow properties and thus compare functionals of single trajectories. This is generally not possible with the master equation approach.

For later use we conclude the section by looking at some prototypical models. A simple analysis shows, somewhat surprisingly, that an innocent-looking example produces second moments that grow indefinitely.

2.1. Reaction Networks and the Master Equation

We consider a chemical network consisting of D different chemical species interacting according to R prescribed reaction pathways. At any given time t, the state of the system is an integer vector $X(t)\in\mathbf{Z}_{+}^{D}=\{0,1,2,\ldots\}^{D}$ counting the number of individual molecules of each species. A reaction law is a prescribed change of state with an intensity defined by a reaction propensity, $w_r:\mathbf{Z}_{+}^{D}\rightarrow\mathbf{R}_{+}$. This is the transition probability per unit of time for moving from the state x to $x-\mathrm{N}_r$:

$$\mathbf{P}\left[X(t+\mathrm{d}t)=x-\mathrm{N}_r\,\middle|\,X(t)=x\right]=w_r(x)\mathrm{d}t+o(\mathrm{d}t). \tag{2.1}$$

where $\mathrm{N}_r\in\mathbf{Z}^{D}$ is the transition step and is the rth column in the stoichiometric matrix $\mathrm{N}\in\mathbf{Z}^{D\times R}$. Informally, for states $x(t)\in\mathbf{R}_{+}^{D}$, we can picture (2.1) as a stochastic version of the time-homogeneous ordinary differential equation

$$x'(t) = -\sum_{r=1}^{R} \mathrm{N}_r w_r(x) = -\mathrm{N}w(x) =: F(x), \tag{2.2}$$

where $w(x) \equiv [w_1(x), \mathrm{L}\ , w_R(x)]^{\mathrm{T}}$ is the column vector of reaction propensities.

The physical premises leading to a description in the form of discrete transition laws (2.1) often imply the existence of a system size V (e.g. physical volume or total number of individuals). For instance, in a given volume V the elementary chemical reactions can be written using the state vector $x = [a, b]^{\mathrm{T}}$,

$$\begin{aligned}
&\varnothing \xrightarrow{k_1 V} A,\ \text{¥}_1 = [-1, 0]^{\mathrm{T}},\\
&A \xrightarrow{k_2 a} \varnothing,\ \text{¥}_2 = [1, 0]^{\mathrm{T}},\\
&A + A \xrightarrow{k_3 a(a-1)/V} \varnothing,\ \text{¥}_3 = [2, 0]^{\mathrm{T}},\\
&A + B \xrightarrow{k_4 ab/V} \varnothing,\ \text{¥}_4 = [1, 1]^{\mathrm{T}},
\end{aligned} \tag{2.3}$$

with the names of the species in capitals. These propensities are generally scaled such that $w_r(x) = V u_r(x/V)$ for some dimensionless function u_r. Intensities of this form are called density dependent and arise naturally in a number of situations ([24] , Chap. 11). For the rest of this paper, we conveniently take $V = 1$ and defer system's size analysis to another occasion.

The models we consider here all have states in the positive integer lattice and the assumption that no transition can yield a state outside $\mathbf{Z}_+^D$ is therefore natural. We make this formal as follows ([25] , Chap. 8.2.2, Definition 2.4):

Assumption 2.1 (Conservation and stability). For all propensities, $w_r(x) = 0$ for any $x \in \mathbf{Z}_+^D$ such that $x - \mathrm{N}_r \notin \mathbf{Z}_+^D$, and we also restrict initial data to $\mathbf{Z}_+^D$. Furthermore, $w_r : \mathbf{Z}_+^D \to \mathbf{R}_+$ such that $w_r(x)$ is finite for all finite arguments x.

To state the chemical master equation (CME), let for brevity $p(x,t) = \mathbf{P}(X(t) = x | X(0) = x_0)$ be the probability that a certain number x of molecules is present at time t conditioned upon an initial state x_0. The CME is then given by ([14] , Chap. V)

$$\frac{\partial p(x,t)}{\partial t} = \sum_{r=1}^{R} w_r(x+\mathrm{N}_r) p(x+\mathrm{N}_r,t) - w_r(x) p(x,t) =: \mathbf{M}^{\mathrm{T}} p(x,t). \tag{2.4}$$

The convention of the transpose of the operator to the right of (2.4) is the standard mathematical formulation of Kolmogorov's forward differential system ([25] , Chap. 8.3) in terms of which $\mathbf{M}$ is the infinitesimal generator of the associated Markov process. This is also the adjoint of the master operator $\mathbf{M}^{\mathrm{T}}$ in the sense that $(\mathbf{M}^{\mathrm{T}} p, q) = (p, \mathbf{M} q)$ in the Euclidean inner product over the state space. An explicit representation is

$$\mathbf{M} q(x) = \sum_{r=1}^{R} w_r(x) \left[q(x-\mathrm{N}_r) - q(x) \right], \tag{2.5}$$

such that the propensities in (2.1) can be retrieved,

$$\mathbf{M}(x, x-\mathrm{N}_r) = w_r(x). \tag{2.6}$$

Under assumptions to be prescribed in Section 4.1 it holds that the dynamics of the expected value of some time-independent unknown function f, conditioned upon the initial state x_0, can be written

$$\begin{aligned} \frac{\mathrm{d}}{\mathrm{d}t} E^{x_0}\left[f(X_t) \right] &= \sum_{x \in Z_+^D} \frac{\partial p(x,t)}{\partial t} f(x) = \left(\mathbf{M}^{\mathrm{T}} p, f \right) = \left(p, \mathbf{M} f \right) \\ &= \sum_{r=1}^{R} E^{x_0}\left[w_r(X_t) \left(f(X_t - \mathrm{N}_r) - f(X_t) \right) \right]. \end{aligned} \tag{2.7}$$

We now consider a path-wise representation for the stochastic process X_t.

2.2. The Sample Path Representation

In the present context of analyzing models in stochastic chemical kinetics, the path-wise jump SDE representation seems to have been first put to use in [26] , and it was later further detailed in [16] . It should be noted, however, that an equivalent representation was used much earlier by Kurtz (see the monograph [24]).

We thus assume the existence of a probability space $(\Omega, \mathrm{F}, \mathbf{P})$ with the filtration $\mathrm{F}_{t\geq 0}$ containing R- dimensional Poisson processes. The state of the system $X(t) \in \mathbf{Z}_+^D$ will be constructed from a stochastic integral with respect to suitably chosen Poisson random measures.

The transition probability (2.1) defines a counting process $\pi_r(t)$ counting at time t the number of reactions of type r that has occurred since $t = 0$. It follows that these processes fully determine the state $X(t)$,

$$X_t = X_0 - \sum_{r=1}^{R} ¥_r \pi_r(t). \tag{2.8}$$

The counting processes are obtained from the transition intensities (cf. (2.1))

$$\mathbf{P}\left[\pi_r(t+\mathrm{d}t) - \pi_r(t) = 1 \middle| \mathrm{F}_t\right] = w_r(X_{t-})\mathrm{d}t + o(\mathrm{d}t), \tag{2.9}$$

where by $X(t-)$ we mean the value of the process prior to any transitions occurring at time t, and where the little-o notation is understood uniformly with respect to the state variable. Alternatively, using Kurtz's random time change representation ([24] , Chap. 6.2), we can produce the counting process from a standard unit-rate Poisson process Π_r,

$$\pi_r(t) = \Pi_r\left(\int_0^t w_r(X_{s-})\mathrm{d}s\right). \tag{2.10}$$

The marked counting measure ([27] , Chap. VIII) $\mu_r(\mathrm{d}t \times \mathrm{d}z; \omega)$ with $\omega \in \Omega$ defines an increasing sequence of arrival times $\tau_i \in \mathbf{R}_+$ with corresponding "marks" $z_i \in I := [0,1]$ according to some probability distribution which we will take to be uniform. The intensity $m_r(\mathrm{d}t \times \mathrm{d}z)$ of $\mu_r(\mathrm{d}t \times \mathrm{d}z)$ is the Lebesgue measure scaled by the corresponding propensity, $m_r(\mathrm{d}t \times \mathrm{d}z) = w_r(X_{t-})\mathrm{d}t \times \mathrm{d}z$. Using this formalism, (2.8) and (2.10) can be written in the jump SDE form

$$\mathrm{d}X_t = -\int_I \mathbb{N}\mu(\mathrm{d}t \times \mathrm{d}z), \tag{2.11}$$

where $\mu = [\mu_1, \ldots, \mu_R]^{\mathrm{T}}$. Here, the time $\tau - t$ to the arrival of the next reaction of type r is exponentially distributed with intensity $w_r(X_{t-})$. Note that, by virtue of the nature of the propensities, the intensities of the counting processes therefore depend nonlinearly on the state ([27] , Chap. II.3).

Using that the point processes are independent and therefore have no common jump times ([25], Chap. 8.1.3), we can obtain a sometimes more transparent notation in terms of a scalar counting measure. Define for this purpose and for any state x the cumulative intensities

$$W_r(x) = \sum_{s=1}^{r} w_s(x), \tag{2.12}$$

such that the total intensity is given by $W(x) \equiv W_R(x)$. Let the marks z_i be uniformly distributed on I. Then the frequency of each reaction can be controlled through a set of indicator functions $\hat{w}_r : \mathbf{Z}_+^D \times I \to \{0,1\}$ defined according to

$$\hat{w}_r(x;z) = \begin{cases} 1 & \text{if } W_{r-1}(x) < zW(x) \le W_r(x), \\ 0 & \text{otherwise.} \end{cases} \tag{2.13}$$

Put $\hat{w}(x) \equiv [\hat{w}_1(x;z), \ldots, \hat{w}_R(x;z)]^{\mathrm{T}}$ and define also for later use the indicator form

$$\hat{F}(x;z) = -\mathbb{N}\hat{w}(x;z), \tag{2.14}$$

such that

$$F(x)=\int_I \hat{F}(x;z)W(x)\mathrm{d}z, \tag{2.15}$$

where $F(x)$ is defined in (2.2).

The jump SDE (2.11) can now be written in terms of a scalar counting random measure μ through a state- dependent thinning procedure ([28] , Chap. 7.5),

$$\mathrm{d}X_t=-\int_I \mathrm{N}\hat{w}(X_{t-};z)\mu(\mathrm{d}t\times\mathrm{d}z). \tag{2.16}$$

Equation (2.16) expresses exponentially distributed reaction times that arrive according to a point process of intensity $m(\mathrm{d}t\times\mathrm{d}z)=W(X_{t-})\mathrm{d}t\times\mathrm{d}z$ carrying a mark which is uniformly distributed in I. This mark implies the ignition of one of the reaction channels according to the acceptance-rejection rule (2.13).

One frequently decomposes (2.16) into its “drift” and “jump” parts,

$$\mathrm{d}X_t=-\mathrm{N}w(X_t)\mathrm{d}t-\int_I \mathrm{N}\hat{w}(X_{t-};z)(\mu-m)(\mathrm{d}t\times\mathrm{d}z). \tag{2.17}$$

The second term in (2.17) is driven by the compensated measure $(\mu-m)$ and is a local martingale provided in essence that the path is absolutely integrable (see [27], Chap. VIII.1, Corollary C4 for details).

2.2.1. Localization: Itô’s and Dynkin’s Formulas

In analytic work it is often necessary to “tame” the process by deriving results under a stopping time $\tau_P := \inf_{t\geq 0}\{\|X_t\|>P\}$ in some norm. Results for the stopped process $X_{t\wedge\tau_P}$ can then be transferred to the original process by letting $P\to\infty$ under suitable conditions.

Although there are many general versions of Itô’s change of variables formula available in the setting of semi- martingales (see for example

[8], Chap. 2.7 and [6] , Chap. II.7), we shall get around with the following simple version ([7] , Chap. 4.4.2). By the properties of the semi-martingale pure jump process we have for $\hat{t} = t \wedge \tau_P$

$$f(X_{\hat{t}}) - f(X_0) = \sum_{0<s\leq\hat{t}} f(X_s) - f(X_{s-}) = \int_0^{\hat{t}}\int_\Gamma f(X_s) - f(X_{s-})\mu(\mathrm{d}s \times \mathrm{d}z), \tag{2.18}$$

where the sum is over jump times $s \in (0, \hat{t}]$. Using that $X_s = X_{s-} - ¥\hat{w}(X_{s-};z)$ we can write this in differential form as

$$\mathrm{d}f(X_t) = \int_\Gamma f(X_{t-} - \mathrm{N}\hat{w}(X_{t-};z)) - f(X_{t-})\mu(\mathrm{d}t \times \mathrm{d}z). \tag{2.19}$$

Alternatively, decomposing (2.18) into drift- and jump parts and taking expectation values we get, since the compensated measure is a local martingale,

$$\begin{aligned} Ef(X_{\hat{t}}) - Ef(X_0) &= E\int_0^{\hat{t}}\int_\Gamma f(X_{s-} - \mathrm{N}\hat{w}(X_{s-};z)) - f(X_{s-})m(ds \times \mathrm{d}z) \\ &= E\int_0^{\hat{t}}\int_I \left[f(X_{s-} - \mathrm{N}\hat{w}(X_{s-};z)) - f(X_{s-})\right]W(X_{s-})\mathrm{d}s \times \mathrm{d}z \\ &= E\int_0^{\hat{t}}\sum_{r=1}^{R}\left[(f(X_s - \mathrm{N}_r) - f(X_s))w_r(X_s)\right]\mathrm{d}s. \end{aligned} \tag{18}$$

This is Dynkin's formula ([25] , Chap. 9.2.2) for the stopped process and we note that (2.7) is just a differential version.

2.2.2. Coupled Processes

When considering stability properties we will need to compare different trajectories with respect to the same noise. The details of this coupling are not defined in either (2.11) or (2.16) and must in fact be chosen explicitly. Since this equality is easy to inspect for a unit-rate Poisson process, the viewpoint of local time expressed in (2.10) provides an answer; two processes X_t and Y_t may be regarded as coupled if and only if they are evolved using identical Poisson processes Π_r, $r = 1, \mathrm{L}, R$ in (2.8) and (2.10). This approach was first used by Kurtz [29] in the context of the random time change representation. Algorithmically it implies the

Common Reaction Path (CRP) method for simulating coupled processes [30] (see also [17]).

A refinement of this construction was devised, also by Kurtz, in ([31] , see Equations (2.2), (2.3)). In turn, this approach implies the Coupled Finite Difference method [18] (but see also [16] [26]), and is more amenable to analysis. This is also the construction formalized below under our current framework.

To obtain such a coupled version of (2.16) we will have to make the thinning dependent on both trajectories. This is achieved by firstly replacing the cumulative intensities in (2.12) with the base (or minimal) intensities

$$W_r^{(0)}(x,y)=\sum_{s=1}^{r} w_s(x)\wedge w_s(y), \tag{2.21}$$

and use the new total base intensity $W^{(0)}(x,y)\equiv W_R^{(0)}(x,y)$ as the intensity of the counting measure μ_0; $m_0(\mathrm{d}t\times \mathrm{d}z)=W^{(0)}(X_{t-},Y_{t-})\mathrm{d}t\times \mathrm{d}z$. We also modify (2.13) accordingly,

$$\hat{w}_r^0(x,y;z)=\begin{cases}1 & \text{if } W_{r-1}^{(0)}(x,y)<zW^{(0)}(x,y)\le W_r^{(0)}(x,y),\\ 0 & \text{otherwise.}\end{cases} \tag{2.22}$$

Secondly, we also define the remainder intensity,

$$W_r^{(\delta)}(x,y)=\sum_{s=1}^{r} w_s(x)\vee w_s(y)-W_r^{(0)}(x,y)=\sum_{s=1}^{r}\left|w_s(x)-w_s(y)\right|. \tag{2.23}$$

In analogy with the previous construction we have the associated total intensity $W^{(\delta)}(x,y)\equiv W_R^{(\delta)}(x,y)$ and counting measure μ_δ; $m_\delta(\mathrm{d}t\times \mathrm{d}z)=W^{(\delta)}(X_{t-},Y_{t-})\mathrm{d}t\times \mathrm{d}z$. This time the thinning procedure is non-symmetric in its two first arguments,

$$\hat{w}_r^\delta(x,y;z)=\begin{cases}1 & \text{if } W_{r-1}^{(\delta)}(x,y)<zW^{(\delta)}(x,y)\le W_{r-1}^{(\delta)}(x,y)+d(x,y),\\ 0 & \text{otherwise,}\end{cases} \tag{2.24}$$

with the non-symmetricity due to

$$d(x,y)=w_r(x)-w_r(x)\wedge w_r(y). \tag{2.25}$$

As a concrete example of how this comparative thinning might be used, consider the following variant of (2.19),

$$df(X_t-Y_t)=\int_r f\left(X_{t-}-\mathrm{N}\hat{w}^{\delta}(X_{t-},Y_{t-};z)-Y_{t-}+\mathrm{N}\hat{w}^{\delta}(Y_{t-},X_{t-};z)\right)-f(X_{t-}-Y_{t-})\mu_{\delta}(\mathrm{d}t\times\mathrm{d}z). \tag{2.26}$$

For this specific example, the terms governed by the base counting measure μ_0 cancel out altogether.

We mention also that an equivalent construction, but one that leads to different algorithms, can be obtained via a thinning of a single measure [16] [26] . Defining instead

$$W_r^{(+)}(x,y)=\sum_{s=1}^{r}w_s(x)\vee w_s(y), \tag{2.27}$$

implying the total intensity $W^{(+)}(x,y)\equiv W_R^{(+)}(x,y)$ and associated counting measure μ_+; $m_+(\mathrm{d}t\times\mathrm{d}z)=W^{(+)}(X_{t-},Y_{t-})\mathrm{d}t\times\mathrm{d}z$. By construction the indicator functions are now non-symmetric in their first two arguments,

$$\hat{w}_r^+(x,y;z)=\begin{cases}1 & \text{if } W_{r-1}^{(+)}(x,y)<zW^{(+)}(x,y)\le W_{r-1}^{(+)}(x,y)+w_r(x),\\ 0 & \text{otherwise.}\end{cases} \tag{2.28}$$

In analogy to (2.26) we get

$$df(X_t-Y_t)=\int_r f\left(X_{t-}-\mathrm{N}\hat{w}^{+}(X_{t-},Y_{t-};z)-Y_{t-}+\mathrm{N}\hat{w}^{+}(Y_{t-},X_{t-};z)\right)-f(X_{t-}-Y_{t-})\mu_{+}(\mathrm{d}t\times\mathrm{d}z). \tag{2.29}$$

This time, however, the intensity of the counting measure is generally larger and the equivalence is obtained as a result of the thinning procedure.

2.2.3. The Validity of the Master Equation

With this much formalism developed, we may conveniently quote the following result:

Theorem 2.1 ([25] , Chap. 8.3.2, Theorem 3.3). Under Assumption 2.1, and if additionally, for $t \in [0,T]$ it holds that

$$EW(X_t) < \infty, \tag{2.30}$$

then (2.7) is valid for $t \in [0,T]$.

Since the governing Equation (2.7) for the expected value of $f(X_t)$ is a direct consequence of (2.4), we can similarly conclude the following:

Corollary 2.2 Under the assumptions of Theorem 2.1, and if, moreover, in an arbitrary norm $\|\cdot\|$,

$$E\|f(X_t)\| < \infty, \quad t \in [0,T], \tag{2.31}$$

then (2.7) is valid for $t \in [0,T]$.

In stating these results we have suppressed the conditional dependency on the initial state which we for simplicity consider to be some non-random state x_0.

2.3. Concrete Examples

Consider the bi-molecular birth-death system,

$$\left.\begin{array}{l} \varnothing \xrightarrow{k_1} A \\ \varnothing \xrightarrow{k_1} B \\ A + B \xrightarrow{k_2 ab} \varnothing \end{array}\right\}, \tag{2.32}$$

that is, the system is in contact with a large reservoir such that A- and B-molecules are emitted at a constant rate k_1. Additionally, a decay reaction happens with probability k_2 per unit of time whenever two molecules meet. For this example we have the stoichiometric matrix

$$\mathrm{N} = \begin{bmatrix} -1 & 0 & 1 \\ 0 & -1 & 1 \end{bmatrix}$$

and the vector propensity function

$$w(x) = [k_1, k_1, k_2 ab]^{\mathrm{T}},$$

where $x = [x_1, x_2]^{\mathrm{T}} = [a, b]^{\mathrm{T}}$.

For a state $X_t = [A_t, B_t]^{\mathrm{T}}$, define $U_t = A_t - B_t \in \mathbf{Z}$. Itô's formula (2.19) with $f(x) = x_1 - x_2$ yields

$$dU_t = df(X_t) = \int_l -[-1,1,0]\hat{w}(X_{t-}; z)\mu(dt \times dz), \tag{2.33}$$

which upon a moments consideration is just the same thing as the model

$$\varnothing \underset{k_1}{\overset{k_1}{f}} U, \tag{2.34}$$

that is, a constant intensity discrete random walk process. An explicit solution is the difference between two independent Poisson distributions,

$$U_t = U_0 + \Pi_1(k_1 t) - \Pi_2(k_1 t) : \; \mathrm{N}\left(U_0, 2k_1 t\right), \quad \text{as } t \to \infty, \tag{2.35}$$

where N is a normally distributed random variable of the indicated mean and variance. Hence U_t fluctuates between arbitrarily large and small values as $t \to \infty$.

Reversible Versions

From time to time below we shall be concerned with the following closed version of (2.32), consisting of a single reversible reaction,

$$A + B \underset{k_2 c}{\overset{k_1 ab}{\rightleftharpoons}} C \tag{2.36a}$$

This is clearly a finite system since the number $a+b+2c$ is always preserved. An open version of the same system is

$$A + B \underset{k_2 c}{\overset{k_1 ab}{\rightleftharpoons}} C \underset{k_4}{\overset{k_3 c}{\rightleftharpoons}} \varnothing, \tag{2.36b}$$

and will prove to be a useful example in the stochastic setting since formally, all states in $\mathbf{Z}_+^3$ are reachable. For (2.36a) we have

$$\mathrm{N} \equiv \begin{bmatrix} 1 & -1 \\ 1 & -1 \\ -1 & 1 \end{bmatrix}, \quad w(x) \equiv [k_1 ab, k_2 c]^{\mathrm{T}}, \tag{2.37}$$

while (2.36b) is represented by

$$\mathrm{N} \equiv \begin{bmatrix} 1 & -1 & 0 & 0 \\ 1 & -1 & 0 & 0 \\ -1 & 1 & 1 & -1 \end{bmatrix}, \quad w(x) \equiv [k_1 ab, k_2 c, k_3 c, k_4]^{\mathrm{T}}. \tag{2.38}$$

These examples, while very simple to deal with, will provide good counterexamples in both Sections 3 and 4.

3. DETERMINISTIC STABILITY

In this section we shall be concerned with the deterministic drift part of the dynamics (2.17). We are interested in techniques for judging the stability of the time-homogeneous ODE (2.2), the so-called reaction rate equations implied by the rates (2.1). Stability and continuity with respect to initial data are considered in Sections 3.1 and 3.2. The main motivation for this discussion stems from the observation that assumptions that do not hold in this very basic setting are unlikely to hold in the stochastic case. In Section 3.3, techniques for explicitly obtaining all our postulated constants are discussed. A good point in favor of taking the time to describe these techniques is that we have not found such a discussion elsewhere.

Initially we will consider states $x \in \mathbf{R}^D$, but we will soon find it convenient to restrict the treatment to $x \in \mathbf{R}_+^D$. In order to remain valid also in the discrete stochastic setting, however, constructed counterexamples will remain relevant also when restricted to $\mathbf{Z}_+^D$.

3.1. Stability

Many stability proofs can be thought of as comparisons with relevant linear cases. This is the motivation for the well-known Grönwall's inequality which we state in the following two versions.

Lemma 3.1. Suppose that $u'(t) \le A + \alpha u(t)$ for $t \ge 0$. Then

$$u(t) \le u(0)\mathrm{e}^{\alpha t} + \frac{A}{\alpha}\left(\mathrm{e}^{\alpha t} - 1\right). \tag{3.1}$$

The same conclusion holds irrespective of the differentiability of u but with $\alpha \ge 0$ and under the weaker integral condition

$$u(t) \le u(0) + \int_0^t A + \alpha u(s)\,\mathrm{d}s. \tag{3.2}$$

The most immediate way of comparing the growth of solutions to the ODE (2.2) to those of a linear ODE is to require that the norm of the driving function is bounded in terms of its argument;

$$\|F(x)\| \le A + \alpha\|x\|, \tag{3.3}$$

since then by the triangle inequality,

$$\|x(t)\| \le \|x_0\| + \int_0^t \|F(x)\| \, dt \le \|x_0\| + \int_0^t A + \alpha \|x(t)\| \, dt, \tag{3.4}$$

where Grönwall's inequality applies. Unfortunately, (3.3) is a too strict requirement for our applications.

Proposition 3.2 The bi-molecular birth-death system (2.32) does not satisfy (3.3).

Proof. We compute $\|F(x)\| = \|-\mathrm{N}w(x)\| = \sqrt{2}|k_1 - k_2 ab|$ for a state $x = [a,b]^T$. Hence for a = b = N = 0, 1,⋯ we have for N large enough that $\|F(x)\| = \sqrt{2}k_2 \cdot N^2 - \sqrt{2}k_1$, which can clearly never be bounded linearly in $\|x\| = \sqrt{2}N$.

The problem with the simple condition (3.3) is that it does not take the direction of growth into account; the offending quadratic propensity in (2.32) actually decreases the number of molecules. To deal with this, let $x \in \mathbf{R}^D$ be an arbitrary vector defining an "outward" direction. The length of the component of the driving function along this direction is $(x, F(x))$ and in order not to have x driven too strongly out along this ray we may, in view of Grönwall's inequality, naturally require that $(x/\|x\|, F(x)) \le \text{constant} \times \|x\|$ for $\|x\|$ sufficiently large. Equivalently, for any x,

$$(x, F(x)) \le A + \alpha \|x\|^2, \tag{3.5}$$

from which one deduces

$$\frac{d}{dt}\frac{\|x\|^2}{2} = (x, F(x)) \le A + \alpha \|x\|^2, \tag{3.6}$$

where Grönwall's inequality applies anew. The assumption (3.5) is weaker than (3.3) since the former implies the latter by the Cauchy-Schwarz inequality. Indeed, as in the proof of Proposition 3.2 it is readily checked that for the bi-molecular birth-death system (2.32), we get

$(x,F(x))=k_1(a+b)-k_2(a+b)ab$ which this time readily can be bounded linearly in terms of $\|x\|^2=a^2+b^2$.

Unfortunately, in the case of an infinite state space and strong dependencies between the species the assumption (3.5) is also often unrealistic.

Proposition 3.3 Neither (2.36a) nor (2.36b) admits a bound of the kind (3.5).

Proof. As in the proof of Proposition 3.2 we look at a ray $x^T=(a,b,c)=(N,N,3N)$ parametrized by a non- negative integer N. For (2.36a) we compute $(x,F(x))=(x,-\mathrm{Nw}(x))=(a+b-c)(k_2c-k_1ab)=k_1N^3-3k_2N^2$, which clearly cannot be bounded linearly in $\|x\|^2=11N^2$. The same argument applies also to (2.36b).

This negative result can perhaps best be appreciated as a kind of loss of information about the dependencies between the species in the functional form of the condition (3.5). The number of A- and B-molecules is strongly correlated with the number of C-molecules such that, in fact, in (2.36a) $a+b+2c$ is a preserved quantity. By contrast, in (3.5) the growth of $\|x\|^2$ is estimated from the sum of the growth of the individual elements of x as if they where independent.

A way around this limitation can be found provided that we leave the general case $x\in\mathbf{R}^D$. We therefore specify the discussion to the positive quadrant $x\in\mathbf{R}_+^D$ and assume from now on that it can be shown a priori that the initial data x_0 belongs to this set and that the subsequent trajectory $x(t)$ never leaves it (compare Assumption 2.1).

It then follows that $\|x\|_1=(\mathbf{1},x)=\mathbf{1}^T x$, where $\mathbf{1}$ is the vector of length D containing all ones. This vector also defines a suitable "outward" vector for states $x\in\mathbf{R}_+^D$ since solutions to the ODE (2.2) cannot grow without simultaneously growing also in the direction of $\mathbf{1}$.

Again, in view of Grönwall's inequality Lemma 3.1, we tentatively require that $(\mathbf{1},F(x)) \le \text{constant} \times \|x\|_1$ for $\|x\|_1$ sufficiently large. Equivalently, for any x,

$$(\mathbf{1},F(x)) \le A + \alpha \|x\|_1, \tag{3.7}$$

implying the bounded dynamics

$$\frac{\mathrm{d}}{\mathrm{d}t}\|x\|_1 = (\mathbf{1},F(x)) \le A + \alpha \|x\|_1. \tag{3.8}$$

We remark in passing that the criterion (3.7) is sharp in the sense that if the reversed inequality can be shown to be true, then the growth of solutions can be estimated from below.

Example 3.1 As a point in favor of this approach we compute for the bi-molecular birth-death system (2.32), $(\mathbf{1},F(x)) = (\mathbf{1},-\mathbb{N}w(x)) = 2k_1 - 2k_2ab$ which evidently falls under the assumption (3.7) with $(A,\alpha) = (2k_1,0)$. For the reversible case (2.36a) we similarly get $(\mathbf{1},F(x)) = -k_1ab + k_2c$ such that (3.7) applies with $(A,\alpha) = (0,k_2)$. Finally, and in the same fashion, the open case (2.36b) is seen to be covered by letting $(A,\alpha) = (k_4,(k_2-k_3)\vee 0)$.

The chosen "outward" vector $\mathbf{1}$ is by no means special. Clearly, any strictly positive vector l may be used in its place since $\|x\|_1$ and $\|x\|_l := l^{\mathrm{T}}x$ are equivalent norms over $\mathbf{R}_+^D$. This is a general and useful observation as it may be used to discard parts of a system that are closed without any restrictions on the associated propensities.

Example 3.2 For the reversible system (2.36a), we have already noted that $a+b+2c$ is a conserved quantity such that the choice $l=[1,1,2]^{\mathrm{T}}$ yields $\mathrm{d}/\mathrm{d}t\|x\|_l = 0$. The open case (2.36b) also benefits from this weighted norm in that we get $\mathrm{d}/\mathrm{d}t\|x\|_l \le 2k_4$.

Example 3.3 A slightly more involved model reads as follows:

$$\left.\begin{array}{ll} \varnothing \xrightarrow{k_1} A & A+B \xrightarrow{k_2 ab} 3C \\ \varnothing \xrightarrow{k_1} B & C \xrightarrow{k_3 c} \varnothing \end{array}\right\}$$

This example has been constructed such that the quadratic reaction increases $\|x\|_1$ and hence (3.7) does not apply. However, taking $l=[3,3,2]^T$ we get

$$\frac{d}{dt}\|x_t\|_l = 6k_1 - 2k_3 c \le 6k_1.$$

This example hints at a general technique for obtaining suitable candidates for the weight vector l. Simply form the matrix N_2 consisting of the columns of N that are affected by superlinear propensities. If a vector $l>0$ annihilating these propensities exists, it can be found in the null-space of $-N_2^T$, readily available by linear algebra techniques. We omit the details.

3.2. Continuity

For well-posedness of the ODE (2.2) we also need continuity with respect to the initial data. We cannot ask for uniform Lipschitz continuity since $\|F(x)-F(y)\| \le L\|x-y\|$ clearly implies (3.3) which we have already refuted. For the same reason, a uniform one-sided Lipschitz condition $(x-y, F(x)-F(y)) \le \lambda\|x-y\|^2$ cannot be assumed to hold since it implies (3.5). The problem here is the global nature of the estimate and it therefore seems to be reasonable to localize this assumption. For instance, one might ask for

$$(x-y, F(x)-F(y)) \le \lambda_R \|x-y\|^2 \quad \text{whenever } \|x\|, \|y\| \le R, \tag{3.9}$$

presumably with some growth restrictions on λ_R. Although very general, such an analysis is likely to be less informative when it comes to estimating actual constants in later results. We shall therefore consider the following simpler version,

$$(x-y,F(x)-F(y))\leq\left(M+\mu\|x+y\|_1\right)\|x-y\|^2, \tag{3.10}$$

where the form of λ_R has been restricted to better suit the present purposes. Trivially, the norms $\|\cdot\|_1$ and $\|\cdot\|$ are equivalent and hence the specific choice made in (3.10) is just a matter of convenience. Since the idea here is to use a priori bounds on x and y when deriving perturbation bounds, using $\|\cdot\|_1$ (or $\|\cdot\|_l$) is natural.

Theorem 3.4 Suppose that the ODE (2.2) satisfies (3.7) and (3.10) and that initial data $x_0\in\mathbf{R}_+^D$ implies a solution $x(t)\in\mathbf{R}_+^D$. Then for any $t\in[0,T]$ there is a unique such solution $x(t)$. Moreover, define $C(t;x_0,y_0)\equiv M+\mu\|x_t+y_t\|_1$, where x_t and y_t are two trajectories associated with initial data $x_0,y_0\in\mathbf{R}_+^D$, respectively. Then

$$\begin{aligned}\|x(t)-y(t)\|&\leq\|x_0-y_0\|\exp\left(\int_0^t C(s;x_0,y_0)\,\mathrm{d}s\right)\\&=\|x_0-y_0\|\left[1+\left(M+\mu\left(\|x_0+y_0\|_1\right)\right)t+O\left(t^2\right)\right].\end{aligned} \tag{3.11}$$

Proof. Combining (3.7) with Grönwall's inequality we get the a priori estimate

$$\|x(t)+y(t)\|_1\leq\|x_0+y_0\|_1+\left(\alpha\|x_0+y_0\|_1+2A\right)\left(e^{\alpha t}-1\right)/\alpha.$$

Hence the (bounded) solution to

$$\frac{\mathrm{d}}{\mathrm{d}t}\|x-y\|^2=2(x-y,F(x)-F(y))\leq 2C(t;x_0,y_0)\|x-y\|^2.$$

is readily found through its integrating factor. The order estimate is a consequence of the fact that

$$\int_0^t \|x_s\|_1 - \|x_0\|_1 \, ds = O\left(t^2\right),$$

since the trajectory is continuous.

3.3. Bounds for Elementary Reactions

As briefly discussed by the end of Section 3.1, finding bounds on A and α in (3.7) as well as a suitable weight-vector l amounts to basic inequalities and some fairly straightforward linear algebra manipulations. In this section we therefore consider precise bounds in (3.10) for the elementary propensities (2.3). Since (3.10) is linear in F, a reasonable approach is to consider linear and quadratic propensities separate (constant propensities trivially satisfy (3.10) with $M = \mu = 0$).

Proposition 3.5 (linear case) Write a set of R linear propensities as $w_r(x) = q_r^T x$, $r = 1, \ldots, R$, each with the corresponding stoichiometric vector N_r. Then $F(x) := -\sum_r \mathrm{N}_r w_r(x)$ satisfies $(x-y, F(x)-F(y)) \le M\|x-y\|^2$ with $M = \mu_2[-\mathrm{N}Q^T]$ in terms of the Euclidean logarithmic norm $\mu_2[\cdot]$ and the matrix Q containing the vectors q_r as columns. In particular, in the case of a single linear propensity and, if as is usually the case, $q_j = k\delta_{jn}$ is all-zero except for a single rate constant k in the nth position, then this reduces to $M = k(-\mathrm{N}_{r,n} + \|\mathrm{N}_r\|)/2$.

Proof. The first assertion is immediate since the smallest such constant M by definition is the logarithmic norm (see e.g. [32]). To compute $\mu_2[-\mathrm{N}_r q_r^T]$ when q has the form indicated, we determine the extremal eigenvalue of $-(\mathrm{N}_r q_r^T + q_r \mathrm{N}_r^T)/2$. By the (signed) scaling invariance of the logarithmic norm we may without loss of generality take $k = 1$. The spectral relation for an eigenpair (λ, z) can be written as

$$-\frac{1}{2}\mathrm{N}_{r,j} z_n = \lambda z_j, \quad j = 1, 2, \ldots, D, \, j \ne n,$$

$$-\frac{1}{2}\mathrm{N}_{r,n}z_n - \frac{1}{2}\mathrm{N}_r^{\mathrm{T}}z = \lambda z_n.$$

For non-zero λ the first relation can be solved for z_j. When inserted into the second relation, using that $z_n \neq 0$ (or otherwise $z = 0$), we get a quadratic equation for λ with a single extremal root.

Example 3.4 The simple special case in Proposition 3.5 is generally sharp except for when there are linear reactions affecting all species considered in the model. For example, in a one-dimensional state space, the single decay $A \to \varnothing$ with propensity $w_1(a) = ka$ allows the optimal value $M = -k$. In general D-dimensional space, a chain with unit rate constants of the form $A_1 \to A_2 \to \mathrm{L} \to A_D \to \varnothing$, or a closed loop in which the last transition is replaced with $A_D \to A_1$, both admit bounds $M \leq 0$ as an inspection of the Gershgorin-discs of $-(\mathrm{N}+\mathrm{N}^{\mathrm{T}})/2$ shows.

Other than for those special examples, for the most important linear cases, Table 1 summarizes the bounds as obtained from the special case in Proposition 3.5 (with all reaction constants normalized to unity).

Proposition 3.6 (quadratic case). Write a general quadratic propensity as $w_r(x) = x^{\mathrm{T}}Sx$ with S a symmetric matrix. Then $F_r(x) = -\mathrm{N}_r w_r(x)$ satisfies (3.10) with $M = 0$ and $\mu = \|x+y\|_1^{-1}\mu_2\left[-\mathrm{N}_r(x+y)^{\mathrm{T}}S\right] \leq \|\mathrm{N}_r\|\|S\|$. For the special case that $S_{ij} = k(\delta_{im}\delta_{jn} + \delta_{jm}\delta_{in})/2$ there holds $\mu \leq k\max_{j\in\{m,n\}}(-\mathrm{N}_{r,j} + \|\mathrm{N}_r\|)/4$.

Proof. Since S is symmetric we have $x^{\mathrm{T}}Sx - y^{\mathrm{T}}Sy = (x+y)^{\mathrm{T}}S(x-y)$. Hence an explicit expression for μ is obtained as follows:

$$\mu \leq \sup_{x,y\in\mathbf{Z}_+^D} \|x+y\|_1^{-1}\|x-y\|^{-2}\left(x-y, -\mathrm{N}_r(x+y)^{\mathrm{T}}S(x-y)\right)$$

$$= \sup_{w\in\mathbf{Z}_+^D}\sup_{v\in\mathbf{Z}^D}\|w\|_1^{-1}\|v\|^{-2}\left(v, -\mathrm{N}_r w^{\mathrm{T}}Sv\right) \leq \sup_{\substack{u\geq 0\\ \|u\|_1\leq 1}}\mu_2\left[-\mathrm{N}_r u^{\mathrm{T}}S\right].$$

The indicated upper bound is derived from the fact that $|\mu_2[\cdot]| \le \|\cdot\|$ [32] . For the useful special case, define first the vector $q = q_1 + q_2$ in terms of $q_{1,j} = k\delta_{jm}(x_n + y_n)/2$, and $q_{2,j} = k\delta_{jn}(x_m + y_m)/2$. Using the fact that the logarithmic norm is sub-additive we can reuse the calculation in the proof of Proposition 3.5.

Example 3.5 The most important quadratic cases are summarized in Table 2. For the dimerizations in the lower half of the table there is also a linear part M in (3.10).

Example 3.6 The bi-molecular birth-death model (2.32) admits the constants $(M,\mu) = \left(0, k_2\left(\sqrt{2}-1\right)/4\right)$ in (3.10). Similarly, the reversible cases (2.36a) and (2.36b) both obeys (3.10) with $(M,\mu) = \left(\sqrt{3}-1\right)/2 \times (k_2, k_1/2)$. All these results are sharp except for the open case (2.36b) for which one can obtain a slightly smaller constant M by using the general formula stated in Proposition 3.5.

Example 3.7 As a highly prototypical example we consider the following natural extension of the bi- molecular birth-death model (2.32),

$$\left.\begin{array}{ll} \emptyset \underset{k_3 a}{\overset{k_1 V}{\rightleftharpoons}} A & \emptyset \underset{k_3 b}{\overset{k_1 V}{\rightleftharpoons}} B \\ A + B \xrightarrow{k_2 ab/V} \emptyset & \end{array}\right\},$$

where in this example it is informative to consider the dependence on the system's size V. It is straightforward to show the bounds $(A,\alpha) \le (2k_1 V, -k_3)$ in (3.7) and hence that the system is effectively bounded despite being of open character. This is seen from the fact that, for states $\|x\|_1 \ge 2k_1/k_3 \cdot V$, the dynamics is dissipative in the $\|\cdot\|_1$-norm. Furthermore, from Proposition 3.5 and 3.6 we get the sharp bounds $(M,\mu) \le \left(-k_1 V, k_2/V \cdot \left(\sqrt{2}-1\right)/4\right)$ in (3.10). It follows that for states $\{x, y\}$ such that $\|x+y\|_1 \gtrsim 9.7 k_1/k_2 \cdot V^2$, the dynamics is contractive in the Euclidean norm. For density dependent propensities we expect that $\|x\| \sim V$ in any norm as V grows, and hence the region of contractivity grows in a relative sense. Intuitively one expects that these results offer an insight into the evolution of the process that is relevant also in the stochastic setting.

4. STOCHASTIC STABILITY

We now consider the properties of the stochastic jump SDE (2.16). For convenience we start by collecting all assumptions in Section 4.1. In the stochastic setting the requirements for existence and uniqueness are slightly stronger than in the deterministic case such that the one-sided bound (3.10) needs to be augmented with an unsigned version, implying essentially the assumption of at most quadratically growing propensities. We demonstrate that this assumption is reasonable by constructing a model involving cubic propensities and with unbounded second moments. On the positive side we show in Section 4.2 that the assumptions are strong enough to guarantee finite moments of any order during finite time intervals.

Table 1. Linear propensities and bounds of M in (3.10).

Reaction	Bound on M
$A \to \varnothing$	0
$A \to B$	$\left(\sqrt{2}-1\right)/2$
$A \to B+C$	$\left(\sqrt{3}-1\right)/2$

We prove existence and uniqueness of solutions to the jump SDE (2.16) in Section 4.3. A sufficient condition for the existence of asymptotic bounds of the pth order moment is given in Section 4.4 where we also derive some stability estimates.

Table 2. Quadratic propensities and bounds of M, μ in (3.10).

Reaction	Bound on μ	M
$A+B\to\varnothing$	$\left(\sqrt{2}-1\right)/4$	
$A+B\to C$	$\left(\sqrt{3}-1\right)/4$	
$A+B\to A$	$1/4$	
$A+B\to A+C$	$\sqrt{2}/4$	
$A+A\to\varnothing$	0	2
$A+A\to B$	$\sqrt{5}/2-1$	$\sqrt{5}/2+1$

4.1. Working Assumptions

We state formally the set of assumptions on the jump SDE (14) as follows.

Assumption 4.9 For arguments x, $y\in\mathbf{Z}_{+}^{D}$, $F(x):=-\mathrm{N}w(x)$, and weighted norm $\|x\|_{l}:=l^{\mathrm{T}}x$ we assume that

1) $-l^{\mathrm{T}}\mathrm{N}w(x)\le A+\alpha\|x\|_{l}$ ("bounded growth"),

2) $\left(-l^{\mathrm{T}}\mathrm{N}\right)^{2}w(x)/2\le B+\beta_{1}\|x\|_{l}+\beta_{2}\|x\|_{l}^{2}$ ("absolutely bounded growth"),

3) $\left(\mathbf{1}^{\mathrm{T}}\mathrm{N}^{2}\right)\left|w(x)-w(y)\right|\le L\left(1+\|x+y\|_{1}\right)\|x-y\|$,

4) $\left(x-y,F(x)-F(y)\right)\le\left(M+\mu\|x+y\|_{1}\right)\|x-y\|^{2}$.

The parameters $\{A, B, \beta_1, \beta_2, L\}$ are assumed to be positive (with β_2 possibly zero) but we allow also negative values of $\{\alpha, M, \mu\}$. The vector l is normalized such that $\min_i l_i = 1$; hence the bound $\|x\|_1 \le \|x\|_l$ is sharp.

After the original draft of the current paper was finished, the author became aware of two other papers discussing very similar conditions [33] [34] . In particular, Assumption 4.1 (1), (2) are also found in ([33] , Condition (1). In fact, these very conditions can be shown to be exactly what is needed to apply the earlier and quite general theory found in ([35] , Theorem 7.1).

In Assumption 4.9 (2) the case $\beta_2 = 0$ will merit special attention. For well-posedness it turns out that we will need to require a higher regularity of the initial data when $\beta_2 > 0$ (see Theorem 4.7) and the condition for ergodicity becomes more restrictive (see Theorem 4.9). In practice, $\beta_2 = 0$ implies that opposing quadratic reactions of the type

$$\left.\begin{array}{l} 2X \xrightarrow{x(x-1)} 3X \\ 2X \xrightarrow{x(x-1)} X \end{array}\right\}, \tag{4.1}$$

are impossible. Similarly, when $\beta_1 = 0$ reactions of the type

$$X \xrightarrow{x} 2X \tag{4.2}$$

are excluded.

Note that (2) and (4) are redundant in the sense that they are both implied by (3). However, as we saw in Section 3.3, in (4) it is often possible to find sharper constants M and μ by considering this bound in isolation. Also, although (3) is stronger than (4), it is in particular valid for quadratic propensities as can be seen from the representation used in the proof of Proposition 3.6,

$$|w_r(x) - w_r(y)| = \left|(x+y)^{\mathrm{T}} S(x-y)\right| \le \|S\| \|x+y\|_1 \|x-y\|. \tag{4.3}$$

The Danger with Cubic Propensities

Assumption 4.1 (2) specifies the discussion to propensities with at most quadratic growth, at least when measured in the direction of the weight vector l. To show that this is natural we now demonstrate that additional care should be taken when considering cubic propensities.

Example 4.2 Consider the model

$$\left.\begin{array}{l} 3X \xrightarrow{x(x-1)(x-2)/2} X \\ 3X \xrightarrow{x(x-1)(x-2)} 4X \end{array}\right\},$$

such that the stoichiometric vector is given by $\mathrm{N}=[-2,1]$, and hence that the drift $-\mathrm{N}w(x)=0$.

Proposition 4.1 For the model in Example 4.2, if $X_0 \geq 3$, then the second moment explodes in finite time.

Proof. Assume that both the second and the third moment are bounded for $t \in [0,T)$ with $T > 0$. From (2.7) we get the governing equation

$$\frac{\mathrm{d}}{\mathrm{d}t} EX_t^2 = E\left[3X_t(X_t-1)(X_t-2)\right],$$

such that the growth of the second moment remains bounded only provided that the third moment remains finite. It is convenient to look at the cumulative third order moment. From (2.7),

$$\frac{\mathrm{d}}{\mathrm{d}t} EC_3(X_t) := \frac{\mathrm{d}}{\mathrm{d}t} EX_t(X_t-1)(X_t-2) = E\left[9C_3(X_t)(X_t-2/3)\right].$$

By the arithmetic-geometric mean inequality, $x-2/3 \geq x-1 \geq \left[x(x-1)(x-2)\right]^{1/3}$, such that by Jensen's inequality,

$$\frac{\mathrm{d}}{\mathrm{d}t} EC_3(X_t) \geq 9E\left[C_3(X_t)^{4/3}\right] \geq 9\left[EC_3(X_t)\right]^{4/3}.$$

We put $u^3 = EC_3(X_t)$ and get the differential inequality

$$\frac{\mathrm{d}(1/u)}{\mathrm{d}t} \le -3,$$

which can be integrated and rearranged to produce the bound

$$E[X_t(X_t-1)(X_t-2)] \ge \frac{X_0(X_0-1)(X_0-2)}{1-3tX_0(X_0-1)(X_0-2)}. \tag{4.4}$$

Hence the third, and consequently also the second moment explode for some finite t whenever $X_0 \ge 3$.

Interestingly, we note that if $X_0 = 3$, then the probability that the cubic decay transition occurs first is $1/3$, and if this happens the state of the system will be stuck with a single molecule indefinitely.

4.2. Moment Bounds

In this section we consider general moment bounds derived from (2.20) using localization. To get some guidance, let us first assume that the differential form of Dynkin's formula (2.7) is valid. Since any trajectory $(X_t)_{t\ge 0}$ by the basic Assumption 2.1 will belong to $\mathbf{Z}_+^D$, we may use that $\|X_t\|_l = (l, X_t)$. Hence from (2.7) with $f(x) = (l, x)$ we get that

$$\frac{\mathrm{d}}{\mathrm{d}t} E\|X_t\|_l = (l, F(X_t)) \le A + \alpha\|X_t\|_l, \tag{4.5}$$

by Assumption 4.1 (1). Clearly, the differential form of Grönwall's inequality in Lemma 3.1 applies here. A correct version of this argument unfortunately looses the sign of α.

Proposition 4.2 If Assumption 4.1 (1) is true, then

$$E\|X_t\|_l \le \|X_0\|_l \exp(\alpha_+ t) + A(\exp(\alpha_+ t) - 1)/\alpha_+,$$

where $\alpha_+ = \alpha \vee 0$.

HerE and below we shall make use of the stopping time $\tau_P = \inf_{t\ge 0}\{\|X_t\|_l > P\}$ and define $\hat{t} = t \wedge \tau_P$.

Proof. From (2.20) with $f(x) = (l, x)$ we get that

$$E\|X_{\hat{t}}\|_l = \|X_0\|_l + E\int_0^{\hat{t}} (l, F(X_s))\mathrm{d}s \le \|X_0\|_l + E\int_0^t A + \alpha_+ \|X_{\hat{s}}\|_l \,\mathrm{d}s, \quad (4.6)$$

By the integral form of Grönwall's inequality in Lemma 3.1 we deduce in terms of $Y_t := X_{t\wedge\tau_P}$ that

$$E\|Y_t\|_l \le \|X_0\|_l \exp(\alpha_+ t) + A(\exp(\alpha_+ t) - 1)/\alpha_+ \quad (4.7)$$

such that the same bound holds for X_t by letting $P \to \infty$.

We attempt a similar treatment for obtaining bounds in mean square. Assuming tactically that (2.7) is valid, writing $\|x\|^2 = x^{\mathrm{T}}x$ we get after some work that

$$\frac{\mathrm{d}}{\mathrm{d}t} E\|X_t\|^2 = E\left[\mathbf{1}^{\mathrm{T}}\mathrm{N}^2 w(X_t) - 2X_t^{\mathrm{T}}\mathrm{N}w(X_t)\right] \quad (4.8)$$

where $\mathrm{N}_{ij}^2 \equiv (\mathrm{N}_{ij})^2$. We expect from Grönwall's inequality that $E\|X_t\|^2$ grows at most exponentially with αt whenever

$$\mathbf{1}^{\mathrm{T}}\mathrm{N}^2 w(x) - 2x^{\mathrm{T}}\mathrm{N}w(x) \le A + \alpha\|x\|^2. \quad (4.9)$$

However, this tentative condition is often violated in practice since the second term $-x^{\mathrm{T}}\mathrm{N}w(x) = (x, F(x))$, and since we already know from

Proposition 3.3 that this quantity does not admit bounds in terms of $\|x\|^2$ even for very simple problems.

More realistic conditions arise when seeking to bound $\|X_t\|_l^2$ instead.

Proposition 4.3 If for some constants γ and C,

$$\left(l^{\mathrm{T}}\mathrm{N}\right)^2 w(x) - 2l^{\mathrm{T}}\mathrm{N}w(x)\|x\|_l \le C + \gamma\|x\|_l^2, \tag{4.10}$$

$(l^{\mathrm{T}}\mathrm{N})^2$ understood elementwise), then $E\|X_t\|_l^2 \le \|X_0\|_l^2 \exp(\gamma_+ t) + C(\exp(\gamma_+ t) - 1)/\gamma_+$

The proof of Proposition 4.3 follows the same pattern as for Proposition 4.2, but using this time $f(x) = \|x\|_l^2 = (l, x)^2$ in (2.20). The condition (4.10) is typically more realistic than (4.9) since we recognize the term $-l^{\mathrm{T}}\mathrm{N}w(x)\|x\|_l = (l, F(x))\|x\|_l$, which under the evidently reasonable Assumption 4.1 (1) is $\le (A + \alpha\|x\|_l)\|x\|_l$. It follows that if $(l^{\mathrm{T}}\mathrm{N})^2 w(x)$ grows at most quadratically with $\|x\|_l$, then this assumption is sufficient to yield bounds in mean square. Stated formally,

Proposition 4.4 Under Assumption 4.1 (1) and (2) the condition (4.10) of Proposition 4.3 is true with $\gamma = 2\beta_2 + 2\alpha + 2$ and $C = 2B + \beta_1^2 + A^2$.

Proof. This is straightforward: we get by the assumptions and Hölder's inequality,

$$\left(l^{\mathrm{T}}\mathrm{N}\right)^2 w(x) - 2l^{\mathrm{T}}\mathrm{N}w(x)\|x\|_l \le 2B + 2\beta_1\|x\|_l + 2\beta_2\|x\|_l^2 + 2\left(A + \alpha\|x\|_l\right)\|x\|_l,$$

where an application of Young's inequality yields the indicated bounds.

As a strong point in favor of our running assumptions we now demonstrate that the above reasoning can be generalized: these two conditions implies finite time stability in any order moment. We note that in a recent manuscript [34], related conditions for the same results are proposed.

Theorem 4.5 (Moment estimate). Under Assumption 4.1 (1) and (2), for any integer $p \geq 1$,

$$E\|X_t\|_l^p \leq \left(\|X_0\|_l^p + 1\right)\exp(Ct) - 1, \tag{4.11}$$

where $C > 0$ is a constant depending on the assumptions and on p.

The proof of Theorem 4.5 and some later results will simplify using the following bound.

Lemma 4.6 Let $H(x) \equiv (x+y)^p - x^p$ with $x \in \mathbf{R}_+$ and $y \in \mathbf{R}$. Then for integer $p \geq 1$ we have the bounds

$$H(x) \leq pyx^{p-1} + 2^{p-4} p(p-1) y^2 \left[x^{p-2} + |y|^{p-2}\right], \tag{4.12}$$

$$|H(x)| \leq p|y| 2^{p-2} \left[x^{p-1} + |y|^{p-1}\right]. \tag{4.13}$$

Proof. Both results follow from Taylor expansions;

$$H(x) = pyx^{p-1} + \frac{p(p-1)}{2} y^2 [x + \theta_1 y]^{p-2},$$

$$|H(x)| = p|y||x + \theta_2 y|^{p-1},$$

respectively, where $\theta_{1,2} \in [0,1]$. Using the triangle inequality and the elementary inequality $(a+b)^p \leq 2^{p-1}(a^p + b^p)$ the lemma is proved.

Proof of Theorem 4.5. Using (2.20) with $f(X_t) \equiv [l^{\mathrm{T}} X_t]^p$ we get

$$E\|X_t\|_l^p = \|X_0\|_l^p + E\int_0^t \underbrace{\sum_{r=1}^R w_r(X_s)\left[\left[l^{\mathrm{T}}(X_s - \mathrm{N}_r)\right]^p - \left[l^{\mathrm{T}} X_s\right]^p\right]}_{\equiv G(X_s)} ds. \tag{4.14}$$

Using Lemma 4.6 (4.12) and Assumption 4.1 (1) and (2) we obtain

$$G(x) \le p\left(A+\alpha\|x\|_l\right)\|x\|_l^{p-1} + 2^{p-3}p(p-1)\left(B+\beta_1\|x\|_l+\beta_2\|x\|_l^2\right)\left(\|x\|_l^{p-2}+\Delta^{p-2}\right),$$

where $\Delta := \|l^T N\|_\infty$. Expanding and using Young's inequality with exponents $\{p/(p-1), p/(p-2)\}$ and conjugate exponents $\{p, p/2\}$,

$$E\|X_{\hat{t}}\|_l^p \le \|X_0\|_l^p + \int_0^t C\left(1+E\|X_{\hat{s}}\|_l^p\right)ds,$$

for some constant C which thus depends on the assumptions. Applying Grönwall's inequality and letting $P \to \infty$ we obtain the stated result.

4.3. Existence and Uniqueness

We shall now prove that the jump SDE (2.16) under Assumption 4.1 has a uniquely defined and locally bounded solution. To this end and following ([8], Section 3.1.2), we introduce the following spaces of path-wise locally bounded processes:

$$S_{\mathrm{F}}^{p,\mathrm{loc}}\left(\mathbf{Z}_+^D\right) = \Big\{X(t,\omega) : X \text{ is } \mathrm{F}_t\text{-adapted and } \mathbf{Z}_+^D\text{-valued such that}$$
$$E\sup_{t\in[0,T]}\|X(t,\omega)\|_1^p < \infty \text{ for } \forall T < \infty\Big\}. \tag{4.15}$$

Theorem 4.7 (Existence). Let X_t be a solution to (2.16) under Assumption 4.1 (1) and (2) with $\beta_2 = 0$.

Then if $\|X_0\|_l^p < \infty$, $\{X_t\}_{t\ge0} \in S_{\mathrm{F}}^{p,\mathrm{loc}}\left(\mathbf{Z}_+^D\right)$. If $\beta_2 > 0$ then the conclusion remains under the additional requirement that $\|X_0\|_l^{p+1} < \infty$.

Proof. Below we let C denote a positive constant which may be different on each occasion used. As before we use the stopping time $\tau_P = \inf_{t\ge0}\{\|X_t\|_l > P\}$ and put $\hat{t} = t \wedge \tau_P$. We get from Itô's formula (with G defined in (4.14))

$$\|X_{\hat{t}}\|_l^p = \|X_0\|_l^p + \int_0^{\hat{t}} G(X_s)\mathrm{d}s + \int_0^{\hat{t}}\int_I \sum_{r=1}^{R} \hat{w}_r(X_{s-};z)\left[\left[l^{\mathrm{T}}(X_{s-}-\mathrm{N}_r)\right]^p - \left[l^{\mathrm{T}}X_{s-}\right]^p\right](\mu - m)(\mathrm{d}s\times\mathrm{d}z).$$

Since the propensities are bounded for bounded arguments (Assumption 2.1), using the stopping time we find that the jump part is absolutely integrable and hence a local martingale $M_{\hat{t}}$. We estimate its quadratic variation under Assumption 4.1 (2),

$$E[M]_{\hat{t}}^{1/2} \le E\left[\int_0^{\hat{t}}\int_I \sum_{r=1}^{R} \hat{w}_r(X_s;z)\left|\left[l^{\mathrm{T}}(X_s-\mathrm{N}_r)\right]^p - \left[l^{\mathrm{T}}X_s\right]^p\right|\mu(\mathrm{d}s\times\mathrm{d}z)\right]$$

$$= E\left[\int_0^{\hat{t}}\int_I \sum_{r=1}^{R} \hat{w}_r(X_s;z)W(X_s)\left|\left[l^{\mathrm{T}}(X_s-\mathrm{N}_r)\right]^p - \left[l^{\mathrm{T}}X_s\right]^p\right|\mathrm{d}s\times\mathrm{d}z\right]$$

$$= E\left[\int_0^{\hat{t}} \sum_{r=1}^{R} w_r(X_s)\left|\left[l^{\mathrm{T}}(X_s-\mathrm{N}_r)\right]^p - \left[l^{\mathrm{T}}X_s\right]^p\right|\mathrm{d}s\right]$$

$$\le E\left[\int_0^{\hat{t}} \sum_{r=1}^{R} p\left|l^T\mathrm{N}_r\right| w_r(X_s)2^{p-2}\left[\|X_s\|_l^{p-1} + \left|l^T\mathrm{N}_r\right|^{p-1}\right]\mathrm{d}s\right] \tag{4.16}$$

$$\le E\left[\int_0^{\hat{t}} C\left(B+\beta_1\|X_s\|_l+\beta_2\|X_s\|_l^2\right)\left[\|X_s\|_l^{p-1}+\Delta^{p-1}\right]\mathrm{d}s\right] \tag{4.17}$$

$$\le E\int_0^{\hat{t}} C\left(1+\|X_s\|_l^p+\beta_2\|X_s\|_l^{p+1}\right)\mathrm{d}s \tag{4.18}$$

where $\Delta = \|l^T\mathrm{N}\|_\infty$. In (4.16) Lemma 4.6 (4.13) was applied and Assumption 4.1 (2) entered in (4.17). Assume first that $\beta_2 = 0$. Then for the drift part we have already constructed a suitable bound in Theorem 4.5 such that

$$\|X_{\hat{t}}\|_l^p \le \|X_0\|_l^p + \int_0^{\hat{t}} C\left(1+\|X_s\|_l^p\right)\mathrm{d}s + |M_{\hat{t}}|.$$

Taking supremum and expectation values we get from Burkholder's inequality ([6] , Chap. IV.4) that

$$E\sup_{s\in[0,\hat{t}]}\|X_s\|_l^p \le \|X_0\|_l^p + \int_0^{\hat{t}} C\left(1+E\sup_{s'\in[0,s]}\|X_{s'}\|_l^p\right)\mathrm{d}s.$$

Writing $\|X\|_t^p(t) \equiv \sup_{s\in[0,t]}\|X_s\|_t^p$ we conclude that

$$E\|X\|_t^p(t\wedge\tau_P) \le \|X_0\|_t^p + \int_0^t C\left(1+E\|X\|_t^p(s\wedge\tau_P)\right)\mathrm{d}s.$$

By Grönwall's inequality we have thus shown that $E\|X\|_t^p(t\wedge\tau_P)$ can be bounded in terms of the initial data and time t. The result now follows by letting $P\to\infty$ and using Fatou's lemma.

Next assume that $\beta_2 > 0$. Then we have to rely more directly on Theorem 4.5 in (4.18),

$$E[M]_{\hat{t}}^{1/2} \le \int_0^{\hat{t}} C\left(1+E\|X_s\|_t^{p+1}\right)\mathrm{d}s \le \left(\mathrm{e}^{C\hat{t}}-1\right)\left(\|X_0\|_t^{p+1}+1\right),$$

where, although there is now a dependency on $\|X_0\|^{p+1}$, the rest of the argument carries through.

For the case that the initial data X_0 is non-deterministic we see that the general quadratic case Assumption 4.1 (2) with $\beta_2 > 0$ requires a one order higher moment of the initial data in order for a solution in $S_{\mathrm{F}}^{p,\mathrm{loc}}\left(\mathbf{Z}_+^D\right)$ to exist.

Theorem 4.8 (Uniqueness) Let Assumption 4.1 (1)-(4) hold true. Then any two paths X_t and Y_t coupled according to the description in Section 2.2.2 with $X_0 = Y_0$ are equal.

We shall be using the observation that, for $x\in\mathbf{Z}_+^D$, we have that $\|x\|_1 \le \|x\|^2$ (referred below to as the "integer inequality").

Proof. Under the same stopping time as before we get from Itô's formula using the coupling described in Section 2.2.2 that

$$\begin{aligned}E\|X_{\hat{t}}-Y_{\hat{t}}\|^2 &= E\int_0^{\hat{t}} 2\left(X_s-Y_s, F(X_s)-F(Y_s)\right)+\left(\mathbf{1}^{\mathrm{T}}\mathrm{N}^2\right)\left|w(X_s)-w(Y_s)\right|\mathrm{d}s \\ &\le E\int_0^{\hat{t}} 2\left(M+\mu\|X_s+Y_s\|_1\right)\|X_s-Y_s\|^2+\left\|\mathbf{1}^{\mathrm{T}}\mathrm{N}^2\right\|_\infty L\left(1+\|X_s+Y_s\|_1\right)\|X_s-Y_s\|\mathrm{d}s.\end{aligned}$$

From the integer inequality we find that there is a constant $C \geq 0$ depending on P such that

$$E\left\|(X-Y)(t\wedge\tau_P)\right\|^2 \leq \int_0^{t\wedge\tau_P} CE\left\|X_s - Y_s\right\|^2 \, ds \leq \int_0^t CE\left\|(X-Y)(s\wedge\tau_P)\right\|^2 \, ds.$$

Using that $E\|X_0 - Y_0\|^2 = 0$ and Grönwall's inequality we conclude that the only solution is the zero solution. Letting $P \to \infty$ and using Fatou's lemma the statement is therefore proved.

In a certain sense the previous result is trivial; from the Poisson representation (2.8) we see that up to the first explosion, a path is uniquely determined from an initial state and a series of Poisson distributed events. However, and as we shall see below, the above proof is prototypical for more involved situations. An example would be when devising hybrid approximations in continuous state space. Indeed, in the above proof, note that if the integer inequality did not hold we would naturally have to rely on the Cauchy-Schwartz inequality instead. With $u(t)^2 = E\|X_t - Y_t\|^2$ this leads to bounds of the typical kind

$$u(t)^2 \leq \int_0^t C\left(u(s) + u(s)^2\right) ds, \tag{4.20}$$

for which $u(t) = \exp(Ct/2) - 1$ is an admissible solution. This observation shows that the integer inequality as used in the proof is crucial; without it the integral inequality (4.19) admits growing solutions.

4.4. Stability

Although Theorem 4.5 shows that any moments are bounded in finite time, a relevant question from the modeling point of view is whether the first few moments remain bounded indefinitely. We give a result to this effect which relies on the existence of solutions in $S_{\mathbb{F}}^{p,\mathrm{loc}}\left(\mathbf{Z}_+^D\right)$ which implies that the differential form (2.7) of Dynkin's formula may be used (cf. Corollary 2.2) such that in turn the differential Grönwall inequality applies. We mention anew that a very similar result has recently appeared in ([33] , Theorem 2).

Theorem 4.9 (Ergodicity) Under Assumption 4.1 (1), (2), suppose that

$$\alpha+\beta_2(p-1)\leq-\kappa_p<0. \tag{4.21}$$

Then for integer $p\geq 1$, $E\|X_t\|_t^p$ remains bounded as $t\to\infty$.

Proof. The case $\beta_2>0$ is slightly more complicated to obtain so we shall concentrate on this. We omit the case $p=1$ since it follows from (4.5) under the present assumptions. The idea of the proof is to asymptotically bound $E(C+\|X_t\|_t)^p$ with a certain positive constant $C=C(p)$ to be decided upon below. By (2.7) we get with $Z_t=C+\|X_t\|_t$,

$$\begin{aligned}\frac{\mathrm{d}EZ_t^p}{\mathrm{d}t}&=E\sum_{r=1}^{R}w_r(X_t)\left[\left[Z_t-l^{\mathrm{T}}\mathrm{N}_r\right]^p-Z_t^p\right]\\&=E\sum_{r=1}^{R}w_r(X_t)\left[-l^{\mathrm{T}}\mathrm{N}_r pZ_t^{p-1}+\frac{\left(-l^{\mathrm{T}}\mathrm{N}_r\right)^2}{2}p(p-1)\left(Z_t-\theta_r l^{\mathrm{T}}\mathrm{N}_r\right)^{p-2}\right],\end{aligned} \tag{4.22}$$

by Taylor's formula for some $\theta_r\in[0,1]$. Using the assumptions we get the bound

$$\frac{\mathrm{d}EZ_t^p}{\mathrm{d}t}\leq pE\left[\left(A+\alpha\|X_t\|_t\right)Z_t^{p-1}+(p-1)\left(B+\beta_1\|X_t\|_t+\beta_2\|X_t\|_t^2\right)\left(Z_t+\left\|l^{\mathrm{T}}\mathrm{N}\right\|_\infty\right)^{p-2}\right]. \tag{4.23}$$

For the first term in (4.23) we get from the scaled Young's inequality with exponent $p/(p-1)$ and conjugate exponent p that

$$\left(A+\alpha\|X_t\|_t\right)Z_t^{p-1}=\alpha Z_t^p+(A-\alpha C)Z_t^{p-1}\leq\alpha Z_t^p+\frac{\varepsilon(p-1)}{p}Z_t^p+\frac{\varepsilon^{1-p}}{p}|A-\alpha C|^p,$$

for some $\varepsilon>0$. As for the second term in (4.23) we first estimate for $\beta_2>0$

$$B+\beta_1\|X_t\|_t+\beta_2\|X_t\|_t^2\leq\beta_2\left(\|X_t\|_t+B/\beta_1\right)\left(\|X_t\|_t+\beta_1/\beta_2\right).$$

Next by the arithmetic-geometric mean inequality we get

$$\left(\|X_t\|_l + B/\beta_1\right)\left(\|X_t\|_l + \beta_1/\beta_2\right)\left(Z_t + \|l^T \mathrm{N}\|_\infty\right)^{p-2}$$
$$\leq \left(\frac{B}{p\beta_1} + \frac{\beta_1}{p\beta_2} + \frac{p-2}{p}\left(\|l^T \mathrm{N}\|_\infty + C\right) + \|X_t\|_l\right)^p = Z_t^p,$$

provided that we choose C as the solution to the equation

$$\frac{B}{p\beta_1} + \frac{\beta_1}{p\beta_2} + \frac{p-2}{p}\left(\|l^T \mathrm{N}\|_\infty + C\right) = C. \tag{4.24}$$

Taken together we thus have

$$\frac{\mathrm{d}EZ_t^p}{\mathrm{d}t} \leq p\left(-\kappa_p + \frac{\grave{o}(p-1)}{p}\right)EZ_t^p + \grave{o}^{1-p}|A - \alpha C|^p.$$

Since $\kappa_p > 0$ we may pick a small enough $\grave{o}$ such that the bracketed expression remains negative. By Grönwall's inequality this then proves the result with $\beta_2 > 0$. To prove the case $\beta_2 = 0$ the same idea of proof applies and results in

$$\frac{\mathrm{d}EZ_t^p}{\mathrm{d}t} \leq p\left(-\kappa_p + \frac{\epsilon(p-1)}{p} + \frac{\epsilon(p-1)^2}{p}\beta_1\right)EZ_t^p + \epsilon^{1-p}\left(|A - \alpha C|^p + \beta_1(p-1)\right).$$

for a certain new constant C satisfying an equation similar to (4.24).

We next aim at deriving some stability estimates with respect to perturbations in the reaction coefficients. An early account of this was given by Kurtz in [31] , see also [18] for a recent discussion in a bounded setting. Given the linear dependence on the coefficients k_r, $r = 1, \ldots, 4$ in the elementary reactions (2.3) a suitable model seems to be that a perturbation $k_r \to k_r + \delta_r$ in a propensity $w_r(x)$ spreads linearly in a relative sense,

$$\left|w_r(x, k_r) - w_r(x, k_r + \delta_r)\right| \leq \text{constant} \times \delta_r w_r(x, k_r). \tag{4.25}$$

We make this formal by requiring that

$$\left(\mathbf{1}^{\mathrm{T}}\mathrm{N}^{2}\right)\left|w(x)-w_{\delta}(x)\right|\leq\delta\left\|w(x)\right\|_{1}\leq\delta C\left(1+\left\|x\right\|_{1}^{2}\right), \tag{4.26}$$

where δ is a suitable measure of the total perturbation vector and where the perturbed propensity vector function is given by

$$w_{\delta}(x)\equiv\left[w_{1}(x,k_{1}+\delta_{1}),\ldots,w_{R}(x,k_{R}+\delta_{R})\right]^{\mathrm{T}}. \tag{4.27}$$

The existence of an absolute constant C in (4.26) follows from Assumption 4.1 (3). We further conveniently assume that the entire statement of Assumption 4.1 carries over to the perturbed system, and for convenience we shall also assume that all constants are the same. By the triangle inequality and Assumption 4.1 (3) we obtain from (4.26) the bound

$$\left(\mathbf{1}^{\mathrm{T}}\mathrm{N}^{2}\right)\left|w(x)-w_{\delta}(y)\right|\leq L\left(1+\left\|x+y\right\|_{1}\right)\left\|x-y\right\|+\delta C\left(1+\left\|y\right\|_{1}^{2}\right)\leq c\left(\delta+\left\|x-y\right\|^{2}\right), \tag{4.28}$$

with C some constant and where the simplification in (4.28) assumes an a priori bound (e.g. stopping time) $\|x+y\|_{1}\leq P$ and additionally requires the integer inequality.

The starting point for the analysis will be Itô's formula under the coupling described in Section 2.2.2. The techniques used below generalize well to pth order moment estimates, but for ease of exposition we let $p=2$.

Hence under the model for coefficient perturbations (4.26)-(4.27) we have that

$$\begin{aligned}\left\|X_{t}-Y_{t}\right\|^{2}&=\int_{0}^{t}2\left(X_{s}-Y_{s},F(X_{s})-F_{\delta}(Y_{s})\right)+\left(\mathbf{1}^{\mathrm{T}}\mathrm{N}^{2}\right)\left|w(X_{s})-w_{\delta}(Y_{s})\right|\mathrm{d}s\\&+\int_{0}^{t}\int_{I}2\left(X_{s-}-Y_{s-},-\mathrm{N}\left(\hat{w}^{\delta}(X_{s-},Y_{s-};z)-\hat{w}^{\delta}(Y_{s-},X_{s-};z)\right)\right)\\&+\left\|-\mathrm{N}\left(\hat{w}^{\delta}(X_{s-},Y_{s-};z)-\hat{w}^{\delta}(Y_{s-},X_{s-};z)\right)\right\|^{2}(\mu_{\delta}-m_{\delta})(\mathrm{d}s\times\mathrm{d}z).\end{aligned} \tag{4.29}$$

Theorem 4.10 (Continuity). Let two trajectories X_{t} and Y_{t} be given, with the same initial data and coupled according to the discussion in Section

2.2.2. Let the propensities for Y_t be perturbed by δ as indi- cated in (4.26), (4.27). Then

$$\lim_{\delta \to 0+} E\|X_t - Y_t\|^2 = 0. \tag{4.30}$$

Proof. We use the stopping time $\tau_P = \inf_{t \geq 0}\{\|X_t + Y_t\|_1 > P\}$ and put $\hat{t} = t \wedge \tau_P$. From (4.29) we get

$$\begin{aligned} E\|X_{\hat{t}} - Y_{\hat{t}}\|^2 &= \int_0^{\hat{t}} E\left[2(X_s - Y_s, F(X_s) - F_\delta(Y_s)) + (\mathbf{1}^T \mathrm{N}^2)|w(X_s) - w_\delta(Y_s)|\right]\mathrm{d}s \\ &\leq \int_0^{\hat{t}} E\left[2(M + \mu\|X_s + Y_s\|_1)\|X_s - Y_s\|^2 + 2\delta(\mathbf{1}^T \mathrm{N}^2)^{1/2}\|w(Y_s) - w_\delta(Y_s)\|\|X_s - Y_s\| + C(\delta + \|X_s - Y_s\|^2)\right]\mathrm{d}s, \end{aligned} \tag{4.31}$$

where (4.28) was used. Simplifying further for a bounded δ we get

$$\begin{aligned} E\|(X - Y)(t \wedge \tau_P)\|^2 &\leq \int_0^{t \wedge \tau_P} C\left(\delta + E\|X_s - Y_s\|^2\right)\mathrm{d}s \\ &\leq \int_0^{t} C\left(\delta + E\|(X - Y)(s \wedge \tau_P)\|^2\right)\mathrm{d}s. \end{aligned}$$

Using that $X_0 = Y_0$ and Grönwall's inequality we conclude that

$$E\|X_{\hat{t}} - Y_{\hat{t}}\|^2 \leq \delta(\exp(Ct) - 1).$$

To get rid of the stopping time we write in terms of indicator functions,

$$\begin{aligned} E\|X_t - Y_t\|^2 &= E\left[\|X_t - Y_t\|^2[t < \tau_P] + \|X_t - Y_t\|^2[t \geq \tau_P]\right] \\ &\leq E\|X_{\hat{t}} - Y_{\hat{t}}\|^2 + \left(E\|X_t - Y_t\|^4\right)^{1/2}(\mathbf{P}[t \geq \tau_P])^{1/2}. \end{aligned} \tag{4.32}$$

Using $\mathbf{P}[t \geq \tau_P] = \mathbf{P}\left[\sup_{s \in [0,t]}\|X_s + Y_s\|_1 > P\right]$ we get from Markov's inequality and the previous estimate

$$E\|X_t - Y_t\|^2 \leq \delta(\exp(Ct) - 1) + \left(E\|X_t - Y_t\|^4\right)^{1/2} P^{-1/2}\left(E\sup_{s \in [0,t]}\|X_s + Y_s\|_1\right)^{1/2}. \tag{4.33}$$

Relying on the existence result in Theorem 4.7 we find that for any given $\varepsilon > 0$ we can select P(and hence also c) such that the right term is $< \varepsilon/2$. We can next find $\delta_0 > 0$ such that for all $\delta \le \delta_0$, also the left term is $< \varepsilon/2$. Hence for all $\delta \le \delta_0$, $E\|X_t - Y_t\|^2 < \varepsilon$ as claimed.

As a by-product of the proof we see that if the process is bounded, then for P large enough the probability in (4.32) is zero.

Corollary 4.11 (Perturbation estimate, bounded version) If in Theorem 4.10, the processes X_t and Y_t are bounded, then for a constant $C > 0$,

$$E\|X_t - Y_t\|^2 \le \delta\left(\exp(Ct) - 1\right). \tag{4.34}$$

The constant C in (4.34) can be bounded explicitly by inspection of (4.28) and (4.31).

For an unbounded system it is apparently much more difficult to obtain explicit estimates. However, by controlling also the martingale part we can strengthen Theorem 4.10 in another direction.

Theorem 4.12 (Continuity/sup) Under the same assumptions as Theorem 4.10 we have that

$$\lim_{\delta \to 0+} E \sup_{s \in [0,t]} \|X_s - Y_s\|^2 = 0. \tag{4.35}$$

Proof. The quadratic variation of the martingale part in (4.29) can be bounded as

$$\begin{aligned}
E[M]_{\hat{t}}^{1/2} &\le E\Big[\int_0^{\hat{t}}\int_I \Big| 2\left(X_{s-} - Y_{s-}, -\mathrm{N}\left(\hat{w}^\delta(X_{s-}, Y_{s-}; z) - \hat{w}^\delta(Y_{s-}, X_{s-}; z)\right)\right) \\
&\quad + \left\|\mathrm{N}\left(\hat{w}^\delta(X_{s-}, Y_{s-}; z) - \hat{w}^\delta(Y_{s-}, X_{s-}; z)\right)\right\|^2 \Big| \mu_\delta(\mathrm{d}s \times \mathrm{d}z)\Big] \\
&\le E\Big[\int_0^{\hat{t}}\int_I \Big(2\|X_s - Y_s\| \left\|\mathrm{N}\left(\hat{w}^\delta(X_s, Y_s; z) - \hat{w}^\delta(Y_s, X_s; z)\right)\right\| \\
&\quad + \left\|\mathrm{N}\left(\hat{w}^\delta(X_s, Y_s; z) - \hat{w}^\delta(Y_s, X_s; z)\right)\right\|^2\Big) W^\delta(X_s, Y_s)\,\mathrm{d}s \times \mathrm{d}z\Big] \\
&= E\Big[\int_0^{\hat{t}} \Big(2\|X_s - Y_s\| \left(\mathbf{1}^T \mathrm{N}^2\right)^{1/2} + \left(\mathbf{1}^T \mathrm{N}^2\right)\Big) \left|w(X_s) - w_\delta(Y_s)\right| \mathrm{d}s\Big] \\
&\le \int_0^{\hat{t}} C\left(\delta + E\|X_s - Y_s\|^2\right) \mathrm{d}s,
\end{aligned}$$

after using (4.28) and the integer inequality anew. For the drift part we may use the corresponding bound developed in the proof of Theorem 4.10. After taking supremum and expectation values of (4.29) and using Burkholder's inequality we therefore arrive at

$$E\|X-Y\|^2(t\wedge\tau_P)\le\int_0^t C\left(\delta+E\|X-Y\|^2(s)\right)\mathrm{d}s\le\int_0^t C\left(\delta+E\|X-Y\|^2(s\wedge\tau_P)\right)\mathrm{d}s\le\delta\left(\exp(Ct)-1\right)$$

by Grönwall's inequality and using the notation $\|X\|(t)\equiv\sup_{s\in[0,t]}\|X_t\|$. We now rely on the same strategy as in the proof of Theorem 4.10 to similarly arrive at

$$E\|X-Y\|^2(t)\le\delta\left(\exp(Ct)-1\right)+\left(E\|X-Y\|^4(t)\right)^{1/2}P^{-1/2}\left(E\|X+Y\|_1(t)\right)^{1/2},$$

and the conclusion follows as before.

5. CONCLUSIONS

We have proposed a theoretical framework consisting of a priori assumptions and estimates for problems in stochastic chemical kinetics. The assumptions are strong enough to guarantee well-posedness for a large and physically relevant class of problems. Long time estimates and limit results for perturbations in rate constants have been studied to exemplify the theory. The assumptions are constructive in the sense that explicit techniques for obtaining all postulated constants have either been worked out in detail or at least indicated. We have seen that the case $\beta_2=0$ in Assumption 4.1 (2) is particularly promising from the analysis point of view in that the conditions for existence in Theorem 4.7 and the ergodicity in Theorem 4.9 both can be formulated naturally.

In the course of motivating our setup we have seen that most problems do not admit global Lipschitz constants and that one-sided versions do not provide a better alternative. Another conclusion worth highlighting is that it pays off to consider jump SDEs in a fully discrete setting in that there are potential complications in proving uniqueness in continuous state space. A practical implication is that care should be exercised when forming continuous approximations to these types of jump SDEs.

For future work we intend to re-visit certain classical results from the perspective of the framework developed herein; for example, thermodynamic limit results, time discretization strategies, and quasi-steady state approximations—all of which have a practical impact in a range of applications.

ACKNOWLEDGEMENTS

The author likes to express his sincere gratitude to Takis Konstantopoulos for several fruitful and clarifying discussions. Early inputs on this work were also gratefully obtained from Henrik Hult, Ingemar Kaj, and Per Lötstedt.

Stefan Engblom This work was supported by the Swedish Research Council within the UPMARC Linnaeus center of Excellence.

REFERENCES

1. Lestas, I., Vinnicombe, G. and Paulsson, J. (2010) Fundamental Limits on the Suppression of Molecular Fluctuations. Nature, 467, 174-178.

2. Raj, A. and van Oudenaarden, A. (2008) Nature, Nurture, or Chance: Stochastic Gene Expression and Its Consequences. Cell, 135, 216-226.

3. Samoilov, M.S. and Arkin, A.P. (2006) Deviant Effects in Molecular Reaction Pathways. Nature Biotechnology, 24, 1235-1240.

4. Gillespie, D.T. (1976) A General Method for Numerically Simulating the Stochastic Time Evolution of Coupled Chemical Reactions. Journal of Computational Physics, 22, 403-434.

5. Gillespie, D.T. (1992) A Rigorous Derivation of the Chemical Master Equation. Physica A, 188, 404-425.

6. Protter, P.E. (2005) Stochastic Integration and Differential Equations. Number 21 in Stochastic Modelling and Applied Probability. 2nd Edition, Version 2.1, Springer, Berlin.

7. Applebaum, D. (2004) Lévy Processes and Stochastic Calculus, Volume 93 of Cambridge Studies in Advanced Mathematics. Cambridge University Press, Cambridge.

8. Situ, R. (2005) Theory of Stochastic Differential Equations with Jumps and Applications. Mathematical and Analytical Techniques with Applications to Engineering. Springer, New York.

9. Øksendal, B. and Zhang, T. (2007) The Itô-Ventzell Formula and Forward Stochastic Differential Equations Driven by Poisson Random Measures. Osaka Journal of Mathematics, 44, 207-230.

10. Hausenblas, E. (2007) SPDEs Driven by Poisson Random Measure with Non Lipschitz Coefficients: Existence Results. Probability Theory and Related Fields, 137, 161-200.

11. Marinelli, C., Prévôt, C. and Röckner, M. (2010) Regular Dependence on Initial Data for Stochastic Evolution Equations with Multiplicative Poisson Noise. Journal of Functional Analysis, 258, 616-649.

12. Marinelli, C. and Röckner, M. (2010) Well-Posedness and Asymptotic Behavior for Stochastic Reaction-Diffusion Equations with Multiplicative Poisson Noise. Electronic Journal of Probability, 15, 1529-1555.

13. Filipović, D., Tappe, S. and Teichmann, J. (2010) Jump-Diffusions in Hilbert Spaces: Existence, Stability and Numerics. Stochastics, 82, 475-520.

14. van Kampen, N.G. (2004) Stochastic Processes in Physics and Chemistry. 2nd Edition, Elsevier, Amsterdam.

15. Higham, D.J. and Kloeden, P.E. (2005) Numerical Methods for Nonlinear Stochastic Differential Equations with Jumps. Numerische Mathematik, 101, 101-119.

16. Li, T.J. (2007) Analysis of Explicit Tau-Leaping Schemes for Simulating Chemically Reacting Systems. Multiscale Modeling & Simulation, 6, 417-436.

17. Engblom, S. (2009) Parallel in Time Simulation of Multiscale Stochastic Chemical Kinetics. Multiscale Modeling & Simulation, 8, 46-68.

18. Anderson, D.F. (2012) An Efficient Finite Difference Method for Parameter Sensitivities of Continuous Time Markov Chains. SIAM Journal on Numerical Analysis, 50, 2237-2258.

19. Kawamura, A. (2009) Lipschitz Continuous Ordinary Differential Equations Are Polynomial-Space Complete. 24th Annual IEEE Conference on Computational Complexity, Paris, 15-18 July 2009, 149-160.

20. Hutzenthaler, M., Jentzen, A. and Kloeden, P.E. (2011) Strong and Weak Divergence in Finite Time of Euler's Method for Stochastic Differential Equations with Non-Globally Lipschitz Continuous Coefficients. Proceedings of the Royal Society A, 467, 1563-1576.

21. Chen, W.Y. and Bokka, S. (2005) Stochastic Modeling of Nonlinear Epidemiology. Journal of Theoretical Biology, 234, 455-470.

22. Ewens, W.J. (2004) Mathematical Population Genetics I. Theoretical Introduction, Volume 27 of Interdisciplinary Applied Mathematics. 2nd Edition, Springer, New York.

23. Escudero, C., Buceta, J., de la Rubia, F.J. and Lindenberg, K. (2004) Extinction in Population Dynamics. Physical Review E, 69, 021908.

24. Ethier, S.N. and Kurtz, T.G. (1986) Markov Processes: Characterization and Convergence. Wiley Series in Probability and Mathematical Statistics. John Wiley & Sons, New York.

25. Brémaud, P. (1999) Markov Chains: Gibbs Fields, Monte Carlo Simulation, and Queues. Number 31 in Texts in Applied Mathematics. Springer, New York.

26. Plyasunov, S. (2005) On Hybrid Simulation Schemes for Stochastic Reaction Dynamics.

27. Brémaud, P. (1981) Point Processes and Queues: Martingale Dynamics. Springer Series in Statistics. Springer, New York.

28. Daley, D.J. and Vere-Jones, D. (2003) An Introduction to the Theory of Point Processes, Volume I: Elementary Theory and Methods. 2nd Edition, Springer, New York.

29. Kurtz, T.G. (1978) Strong Approximation Theorems for Density Dependent Markov Chains. Stochastic Processes and Their Applications, 6, 223-240.

30. Rathinam, M., Sheppard, P.W. and Khammash, M. (2010) Efficient Computation of Parameter Sensitivities of Discrete Stochastic Chemical Reaction Networks. Journal of Chemical Physics, 132, 034103.

31. Kurtz, T.G. (1982) Representation and Approximation of Counting Processes. In: Fleming, W.H. and Gorostiza, L.G., Eds., Advances in Filtering and Optimal Stochastic Control, Vol. 42, Lecture Notes in Control and Information Sciences, Springer, Berlin, 177-191.

32. Ström, T. (1975) On Logarithmic Norms. SIAM Journal on Numerical Analysis, 12, 741-753.

33. Briat, C., Gupta, A. and Khammash, M. (2014) A Scalable Computational Framework for Establishing Long-Term Behavior of Stochastic Reaction Networks. PLoS Computational Biology, 10, e1003669.

34. Rathinam, M. (2014) Moment Growth Bounds on Continuous Time Markov Processes on Non-Negative Integer Lattices. To appear in Quart. Appl. Math.

35. Meyn, S.P. and Tweedie, R.L. (1993) Stability of Markovian Processes III: Foster-Lyapunov Criteria for Continuous- Time Processes. Advances in Applied Probability, 25, 518-548.

CHAPTER 4

Lebesgues-Stieltjes Integrals of Fuzzy Stochastic Processes with Respect to Finite Variation Processes

Jinping Zhang, Lingli Luo, Xingmei Li, Xiaoying Wang

Department of Mathematics and Physics, North China Electric Power University, Beijing, China

ABSTRACT

Let $G=\{G_t(\omega),t\in[0,T]\}$ be a fuzzy stochastic process and $\{A_t(\omega),t\in[0,T]\}$ be a real valued finite variation process. We define the Lebesgue-Stieltjes integral denoted by $\int_0^t G_s(\omega)\mathrm{d}A_s(\omega)$ for each $t>0$ by using the selection method, which is direct, nature and different from the indirect definition appearing in some references. We shall show that this kind of integral is also measurable, continuous in time t and bounded a.s. under the Hausdorff metric.

KEYWORDS

Fuzzy Stochastic Process, Finite Variation Process, Fuzzy Stochastic Lebesgue-Stieltjes Integral, Measurability

1. INTRODUCTION

Recently, the theory of fuzzy functions has been developed quickly due to the measurements of various uncertainties arising not only from the randomness but also from the vagueness in some situations. For example, when considering wave height at time t denoted by f_t, due to the influence of random factors and the limitations of the measurement tools and methods, we may not precisely know the height f_t. It is reasonable to consider the wave height as a fuzzy random variable on a probability space $(\Omega,\mathcal{F},P)$.

Since Puri and Ralescu [1] (1986) defined fuzzy random variable, there had been many further topics such as expectations of fuzzy random variables, fuzzy stochastic processes, integrals of fuzzy stochastic processes, fuzzy stochastic differential equations etc. In order to study a fuzzy function u, it is natural and equivalent to study its α -level set $[u]^\alpha$ for any $\alpha\in[0,1]$, where $[u]^\alpha$ is a set-valued function. Therefore, as usual, in order to explore the integrals of fuzzy stochastic processes, at first we can study the integrals of set-valued stochastic processes. Kisielewicz (1997) [2] used all selections to define the integral of a set-valued process as a nonempty closed subset of $L^2(\Omega,\mathcal{F},P;R^n)$, but did not consider its measurability. Based on Kisielewicz's work (1997) [2] , Kim and Kim (1999) [3] studied some properties of this kind of integral. Jung and Kim (2003) [4] modified the definition in 1-dimensional Euclidean space R so that the integral became a set-valued random variable. After the work [4], there are some references on set-valued integrals and fuzzy integrals. One may refer to papers such as [5] -[13] etc. and references therein. Zhang and Qi [14] (2013) considered the set-valued integral with respect to a finite variation process directly instead of taking the decomposable closure appearing in [4] [6] and other references. As a further work of [14], here we shall explore the integrals of fuzzy stochastic processes with respect to finite variation processes and prove the measurability and boundedness of this kind of integral, the continuity with respect to t under the Hausdorff metric and its representation theorem.

This paper is organized as follows: in Section 2, we present some notions on set-valued random variables and fuzzy set-valued random variables;

in Section 3, we shall give the definition of integral of fuzzy set-valued stochastic processes with respect to finite variation process and prove the measurability and L^2-boundedness which are necessary to our future work on fuzzy stochastic differential equations.

2. PRELIMINARIES

We denote N the set of all natural numbers, R the set of all real numbers, R^d the d-dimensional Euclidean space with the usual norm $\|\cdot\|$, R^+ the set of all nonnegative numbers. Let $(\Omega,\mathcal{F},P)$ be a complete probability space, $\{\mathcal{F}_t, t\in[0,T]\}$ a σ-field filtration satisfying the usual conditions such that $\mathcal{F}_0$ includes all P-null sets in $\mathcal{F}$. The filtration is non-decreasing and right continuous. Let $\mathcal{B}(E)$ be a Borel field of a topological space E.

Let $(\Omega,\mathcal{F},P)$ be a complete probability space. $L^p(\Omega,\mathcal{F},P;R^d)$ (or brief $L^p(\Omega;R^d)$) $(p\geq 1)$ the set of all R^d-valued Borel measurable functions $f:\Omega\to R^d$ such that the norm

$$\|f\|_p=\left\{\int_\Omega\|f(\omega)\|^p\,dP\right\}^{\frac{1}{p}},\ \text{if } 1\leq p<\infty,$$
$$\|f\|_\infty=\operatorname{esssup}_{\omega\in\Omega}\|f(\omega)\|,\ \text{if } p=\infty$$

is finite. f is called L^p-integrable if $f\in L^p(\Omega;R^d)$.

Let $\mathcal{K}(R^d)$ (resp. $\mathcal{K}_k(R^d)$, $\mathcal{K}_{kc}(R^d)$) be the family of all nonempty, closed (resp. nonempty compact, nonempty compact convex) subsets of R^d. For any $x\in R^d$ and $A\in\mathcal{K}(R^d)$, define the distance between x and A by $d(x,A)=\inf_{y\in A}\|x-y\|$. The Hausdorff metric d_H on $\mathcal{K}(R^d)$ (cf. [15]) is defined by

$$d_H(A,B)\doteq\max\left\{\sup_{a\in A}d(a,B),\sup_{b\in B}d(b,A)\right\}\ A,B\in\mathcal{K}(R^d).$$

Denote $\|A\|_K \doteq d_H(\{0\},A) = \sup_{a\in A}\|a\|$.

For $A \subset R^d, x^* \in R^d$, the support function of A is defined as follows:

$$S(x^*,A) = \sup\{\langle x^*,x\rangle : x \in A\}.$$

A set-valued function $F:\Omega \to \mathcal{K}(R^d)$ is said to be measurable if for any open set $O \subset R^d$, the inverse $F^{-1}(O) := \{\omega \in \Omega : F(\omega) \cap O \neq \varnothing\}$ belongs to $\mathcal{F}$. Such a function F is called a set-valued random variable.

Let $\mathcal{M}(\Omega,\mathcal{F},P;\mathcal{K}(R^d))$ (resp. $\mathcal{M}(\Omega,\mathcal{F},P;\mathcal{K}_c(R^d))$, $\mathcal{M}(\Omega,\mathcal{F},P;\mathcal{K}_{kc}(R^d))$) be the family of all measurable $\mathcal{K}(R^d)$-valued (resp. $\mathcal{K}_c(R^d)$, $\mathcal{K}_{kc}(R^d)$-valued) functions, briefly by $\mathcal{M}(\Omega,\mathcal{K}(R^d))$ (resp. $\mathcal{M}(\Omega,\mathcal{F},P;\mathcal{K}_c(R^d))$, $\mathcal{M}(\Omega,\mathcal{F},P;\mathcal{K}_{kc}(R^d))$. For $F \in \mathcal{M}(\Omega,K(R^d))$, the family of all L^p-integrable selections is defined by

$$S_F^p(\mathcal{F}) := \{f \in L^p(\Omega,\mathcal{F},P;R^d) : f(\omega) \in F(\omega)\ a.s.\}, p \geq 1 \qquad (1)$$

In the following, $S_F^p(\mathcal{F})$ is denoted briefly by S_F^p.

A set-valued random variable F is said to be integrable if S_F^1 is nonempty. F is called $L^p(p \geq 1)$-integrably bounded if there exits $h \in L^p(\Omega,\mathcal{F},P;R^d)$ s.t. for all $x \in F(\omega)$, $\|x\| \leq h(\omega)$ almost surely.

An R^d-valued stochastic process $f = \{f_t : t \geq 0\}$ (or denoted by $f = \{f(t) : t \geq 0\}$) is defined as a function $f : R_+ \times \Omega \to R^d$ with the $\mathcal{F}$-measurable section f_t, for $t \geq 0$. We say f is measurable if f is $\mathcal{B}(R_+) \otimes \mathcal{F}$-measurable. The process $f = \{f_t : t \geq 0\}$ is called $\mathcal{F}_t$-adapted if f_t is $\mathcal{F}_t$-measurable for every $t \geq 0$. Let $\Sigma := \bigcap_{t\geq 0}\{Z \in \mathcal{B}(R_+) \otimes \mathcal{F} : Z_t \in \mathcal{F}_t\}$, where $Z_t = \{\omega; (t,\omega) \in Z\}$. We know that Σ is a σ-algebra on $R_+ \times \Omega$. A function $f : R_+ \times \Omega \to R^d$ is measurable and $\mathcal{F}_t$-adapted if and only if it is Σ-measurable ([9]).

In a fashion similar to the R^d-valued stochastic processes, a set-valued stochastic process $F=\{F_t : t \geq 0\}$ is defined as a set-valued function $F: R_+ \times \Omega \to \mathcal{K}(R^d)$ with $\mathcal{F}$-measurable section F_t for $t \geq 0$. It is called measurable if it is $\mathcal{B}_+ \otimes \mathcal{F}$-measurable, and $\mathcal{F}_t$-adapted if for any fixed t, $F_t(\cdot)$ is $\mathcal{F}_t$-measurable. $F=\{F_t : t \geq 0\}$ is measurable and $\mathcal{F}_t$-adapted if and only if it is Σ-measurable. $F=\{F_t : t \geq 0\}$ is called L^p-integrable if every F_t is L^p-integrable.

Let $F(R^d)$ denote the family of all fuzzy sets $v: R^d \to [0,1]$ which satisfy the following two conditions (cf. [3] [6]):

1) The level set $v_1 = \{x \in R^d : v(x)=1\} \neq \emptyset$;

2) Each v is upper semi-continuous function, i.e. for each $\alpha \in (0,1]$, the level set $[v]^\alpha \doteq \{x \in R^d : v(x) \geq \alpha\}$ is a closed subset of R^d.

Let $F_k(R^d)$ denote the family of all fuzzy sets $v: R^d \to [0,1]$ which satisfy the above conditions 1), 2), and

3) The support set $[v]^0 \doteq cl\{x \in R^d : v(x) > 0\}$ is a compact set.

A fuzzy set v is convex if

$$v(\lambda x + (1-\lambda)y) \geq \min\{v(x), v(y)\} \text{ for any } x, y \in R^d, \lambda \in [0,1].$$

It is know that v is convex if and only if, for any $\alpha \in (0,1]$, the level set $[v]^\alpha$ is a convex subset of R^d. Let $F_c(R^d)$ denote the family of all convex fuzzy sets in $F(R^d)$, and $F_{kc}(R^d)$ be the subset of all convex fuzzy sets in $F_k(R^d)$. Define $d_\infty : F(R^d) \times F(R^d) \to [0,\infty)$ (cf. [1]) by the expression

$$d_\infty(u,v) := \sup_{\alpha \in [0,1]} d_H\left([u]^\alpha, [v]^\alpha\right).$$

We know that d_∞ is a metric in $F(R^d)$ and $(F(R^d),d_\infty)$ a complete metric space (cf. [6] [3]). Moreover, for every $u,v,w,z\in F(R^d),\lambda\in R$, we have

$$d_\infty(u+w,v+w)=d_\infty(u,v);$$

$$d_\infty(u+v,w+z)\le d_\infty(u,w)+d_\infty(v,z);$$

$$d_\infty(\lambda u,\lambda v)=|\lambda|d_\infty(u,v).$$

Lemma 1. (cf. [16]) Let B be a set and $\{B_\alpha,\alpha\in[0,1]\}$ be a family of subsets of B such that

1) $B_0=B$;

2) $\alpha\le\beta$ implies $B_\beta\subseteq B_\alpha$;

3) $\alpha_1\le\alpha_2\le\cdots$, $\lim_{n\to\infty}\alpha_n=\alpha$ implies $B_\alpha=\bigcap_{i=1}^{\infty}B_{\alpha_i}$.

Then the function $\phi:B\to[0,1]$ defined by $\phi(x)=\sup\{\alpha\in[0,1]:x\in B_\alpha\}$ has the property that $\{x\in B:\phi(x)\ge\alpha\}=B_\alpha$ for every $\alpha\in[0,1]$.

A mapping $G:\Omega\to F(R^d)$ is said to be measurable if $[G(\omega)]^\alpha:\Omega\to\mathcal{K}(R^d)$ is an set-valued random variable for each $\alpha\in(0,1]$. Such a mapping G is called a fuzzy random variable (cf. [17]). Let $\mathcal{M}(\Omega,\mathcal{F},P;F(R^d))$ (briefly by $\mathcal{M}(\Omega,F(R^d))$) denote the family of all $\mathcal{F}$-measurable fuzzy random variables. As a similar manner, we have the notations $\mathcal{M}(\Omega,\mathcal{F},P;F_k(R^d))$, and $\mathcal{M}(\Omega,\mathcal{F},P;F_{kc}(R^d))$, or briefly by $\mathcal{M}(\Omega,F_k(R^d))$ (resp. $\mathcal{M}(\Omega,F_{kc}(R^d))$).

$G(t,\omega):\Omega\times[0,T]\to F(R^d)$ is called a fuzzy stochastic process if for any $t\in[0,T]$, $G(t,\omega)$ is a fuzzy random variable. A fuzzy stochastic process $G(t,\omega)$ is said to be $\mathcal{F}_t$-adapted, if for every $\alpha\in[0,1]$, the set-valued function $[G(t,\omega)]^\alpha:\Omega\to\mathcal{K}(R^d)$ is $\mathcal{F}_t$-measurable for all $t\in[0,T]$. It is called measurable, if $[G(t,\omega)]^\alpha : \Omega \to \mathcal{K}(R^d)$ is a $\mathcal{B}([0,T]) \otimes \mathcal{F}$-measurable for all $\alpha \in [0,1]$.

A fuzzy stochastic process G is called L^p-integrably bounded, if there exists a real-valued stochastic process $h \in L^p(\Omega \times [0,T], R^d)$, for any $\alpha \in [0,1]$ such that $\|[G(t,\omega)]^\alpha\|_K \le h(t,\omega)$ for any $(t,\omega) \in \Omega \times [0,T]$. It is equivalent to that $\|[G]^0\|_K \in L^p(\Omega \times [0,T], R^d)$.

Let $L^p(\Omega, \mathcal{F}, P; F(R^d))$ denote the family of all measurable $F(R^d)$-valued L^p-integrably bounded fuzzy functions. Write for brevity by $L^p(\Omega, F(R^d))$, where we consider $F, G \in L^p(\Omega, \mathcal{F}, P; F(R^d))$ as identical if $P([F]^\alpha = [G]^\alpha, \forall \alpha \in [0,1]) = 1$. Let $L^p(\Omega \times [0,T], F(R^d))$ denote the family of all L^p-integrably bounded $F(R^d)$-valued $\mathcal{F}_t$-adapted fuzzy stochastic processes.

Let $G : \Omega \to F(R^d)$ be a fuzzy random variable and $p \ge 1$, The following conditions are equivalent (cf. [15]):

1) $G \in L^p(\Omega, \mathcal{F}, P; F(R^d))$;

2) $[G]^0 \in L^p(\Omega, \mathcal{F}, P; \mathcal{K}(R^d))$;

3) $\|[G]^0\|_K \in L^p(\Omega, \mathcal{F}, P; R^+)$.

We define $\hat{\theta} \in F(R^d)$ as $\hat{\theta} = I_{\{0\}}$, where for $x \in R^d$, we have $I_{\{x\}}\{y\} = 1$ if $x = y$ and $I_{\{x\}}\{y\} = 0$ if $x \ne y$.

3. LEBESGUE-STIELTJES INTEGRALS WITH RESPECT TO FINITE VARIATION PROCESSES

Let $(\Omega,\mathcal{F},P)$ be a complete probability space equipped with the usual filtration $\{\mathcal{F}_t,t\in[0,T]\}$. Let $A=\{A_t,t\geq 0\}$ be a real valued $\mathcal{F}_t$-adapted measurable process with finite variation and continuous sample trajectories a.s. from the origin. That is to say, for each compact interval $[s,t]\subset[0,\infty)$ and any partition $\Delta=\{t_1,t_2,\cdots,t_n\}$ of $[s,t]$, the total variation

$$V_A([s,t])=\sup_{\Delta}\sum_{i=1}^{n}\left|A_{t_i}(\omega)-A_{t_{i-1}}(\omega)\right|$$

is finite and $A(0,\cdot)=0$ as. Then for any $T>0$, the process $A=\{A_t,t\geq 0\}$ can generate a random measure denoted by μ_A in the space $([0,T],\mathcal{B}([0,T]))$. For any $(s,t]\subset[0,T]$, let

$$\mu_A((s,t]):=\left|A(t,\omega)\right|-\left|A(s,\omega)\right|$$

where $A(t,\omega)=A^{+}(t,\omega)-A^{-}(t,\omega)$ is the decomposition of A. A^{+} and A^{-} are non-negative and non-de- creasing processes. $\left|A(t,\omega)\right|=A^{+}(t,\omega)+A^{-}(t,\omega)$.

In the product space $(\Omega\times[0,T],\Sigma)$, Michta (2011) in [7] defined a measure as follows:

$$\nu_A(C):=\int_{\Omega}\int_{[0,T]}I_C(t,\omega)\mu_A([0,T])\mu_A(dt)P(d\omega)$$

For $C\in\Sigma$, where I_C is the index function. Then the set function v is a finite measure in the measurable space $(\Omega\times[0,T],\Sigma)$ if and only if $\int_{\Omega}(\mu_A([0,T]))^2 P(d\omega)<\infty$ (cf. [7]). In the following we always assume $\int_{\Omega}(\mu_A([0,T]))^2 P(d\omega)<\infty$.

For $1 \le p < \infty$, let $L^p(\Omega\times[0,T],\Sigma,\nu_A;R^d)$ be the family of all Σ-measurable R^d-valued stochastic processes f such that

$$\int_{\Omega\times[0,T]} \|f(\omega,t)\|^p \, d\nu_A < \infty.$$

For any $f \in L^p(\Omega\times[0,T],\Sigma,\nu_A;R^d)$ and $[s,t]\subset[0,T]$, the stochastic Lebesgue-Stieltjes integral $\int_{[s,t]} f(\tau)\mathrm{d}A_\tau$ is defined by the Bochner integral $\int_{[s,t]} f(\tau)\mu_A(\mathrm{d}\tau)$ path-by-path. One can prove that the integral process $\left\{\int_{[0,t]} f(s)\mathrm{d}A_s, t\in[0,T]\right\}$ is Σ-measurable.

Lemma 2. (cf. [8]) Let $(E,\mathcal{B},\mu)$ be a σ-finite measure space and X a separable Banach space. If $\mathcal{B}$ is separable with respect to μ, (i.e. there exists a countably generated sub-sigma algebra $\mathcal{B}_0 \in \mathcal{B}$ such that for every $A\in\mathcal{B}$, there is $B\in\mathcal{B}_0$ satisfying $\mu(A\Delta B)=0$), then space $L^p(E;X)(p\ge 1)$ is separable in norm.

From now on, we always assume the sigma-field $\mathcal{F}$ is separable with respect to P such that the set-valued integral and fuzzy integral can be well defined.

Let $M^p(\Omega\times[0,T],\Sigma,\nu_A;\mathcal{K}(R^d))$ be the family of all Σ-measurable $\mathcal{K}(R^d)$-valued stochastic processes F such that

$$\int_{\Omega\times[0,T]} \|F(\omega,t)\|_K^p \, d\nu_A < \infty,$$

where $\|F(\omega,t)\|_K = \sup_{x\in F(\omega,t)}\|x\|$.

For any $F \in M^p(\Omega\times[0,T],\Sigma,\nu_A;\mathcal{K}(R^d))$, set

$$S^p(F) := \left\{f \in M^p(\Omega\times[0,T],\Sigma,\nu_A;R^d) : f(\omega,t)\in F(\omega,t), \nu_A - a.e\right\}. \tag{2}$$

Definition 1. (cf. [7]) For a set-valued stochastic process $F \in M^p(\Omega\times[0,T];\mathcal{K}_{kc}(R^d))$ the set-valued stochastic Lebesgue-Stieltjes integral (over interval $[s,t]$) of F with respect to the finite variation continuous process A is the set

$$\int_{[s,t]} F(\tau)\mathrm{d}A_\tau := \left\{\int_{[s,t]} f(\tau)\mathrm{d}A_\tau : f \in S^p(F)\right\}.$$

For some fuzzy stochastic process $G \in \mathcal{M}(\Omega\times[0,T],\mathcal{F}(R^d))$, it is natural to define the fuzzy integral of G with respect to the finite variation process level-wise.

Let $M^p(\Omega\times[0,T],\Sigma,v_A;\mathcal{F}(R^d))$ (or abbrev. as $M^p(\Omega\times[0,T],\mathcal{F}(R^d))$) be the family of all Σ-measu- rable $\mathcal{F}(R^d)$-valued fuzzy stochastic processes G such that

$$\int_{\Omega\times[0,T]} \left\|\left[G(\omega,t)\right]^0\right\|_K^p \mathrm{d}v_A < \infty,$$

where $\left\|\left[G(\omega,t)\right]^0\right\|_K = \sup_{x\in[G(\omega,t)]^0}\|x\|$.

For a fuzzy stochastic process $G \in M^1(\Omega\times[0,T],\mathcal{F}(R^d))$, according to Lemma 1 and the properties of set-valued stochastic integrals, the Lebesgue-Stieltjes integral of G (over interval $[s,t]$) can be defined level-wise.

Set

$$\left[\int_s^t G(\tau,\omega)\mathrm{d}A_\tau(\omega)\right]^\alpha := \int_s^t \left[G(\tau,\omega)\right]^\alpha \mathrm{d}A_\tau(\omega) = \left\{\int_s^t g_\tau(\omega)\mathrm{d}A_\tau(\omega), g \in S^1_{[G]^\alpha}\right\} \tag{3}$$

for all $\alpha \in [0,1]$.

Definition 2. For a fuzzy stochastic process $G \in M^1(\Omega\times[0,T],\mathcal{F}(R^d))$ and any $0 \le s < t \le T$, the family $\left\{\left[\int_s^t G(\tau,\omega)\mathrm{d}A_\tau(\omega)\right]^\alpha, \alpha \in [0,1]\right\}$ defined by Equation (3) can determine an $\mathcal{F}(R^d)$-valued function denoted by $\int_0^t G(s,\omega)\mathrm{d}A_s$, such a fuzzy function is called the Lebesgue-Stieltjes integral (over interval $[s,t]$) of G with respect to finite variation process $A(t)$.

Theorem 1. ([12]) For $F \in M^1(\Omega\times[0,T],\Sigma,K_{kc}(R^d);v_A)$, $[s,t]\subset[0,T]$ and $\omega\in\Omega$, the Lebesgue- Stieltjes integral $\int_s^t F_\tau(\omega)\mathrm{d}A_\tau(\omega)$ is a compact and convex subset of R^d.

Lemma 3. (cf. [18]) Let $(\Omega,\mathcal{F},P)$ be a probability space, X a separable Banach space. For random variables F_1, F_2, both the support function $S(x^*,F(\omega))(x^*\in X^*)$ and the metric $d_H(F_1(\omega),F_2(\omega))$ are $\mathcal{F}$ -measurable.

Lemma 4. (cf. [14]) Let $A_t(\omega)$ be an R-valued stochastic process with finite variation. For $F\in L^2(\Omega\times[0,T],\Sigma,v_A;K_{kc}(R^d))$ and $[s,t]\subset[0,T]$, we have

1) $\int_s^t \alpha F_\tau \mathrm{d}A_\tau = \alpha\int_s^t F_\tau \mathrm{d}A_\tau, \alpha\in[0,1]$;

2) $S\left(x^*,\int_s^t F_\tau \mathrm{d}A_\tau\right)=\int_s^t S\left(x^*,F_\tau\right)\mathrm{d}A_\tau, x^*\in R^d$.

Lemma 5. (cf. [18]) Let $(\Omega,\mathcal{F})$ be a measurable space, X a separable Banach space. Taking $F:\Omega\to\mathcal{K}(X)$ and for any $x^*\in X^*$, assume $S(x^*,F)$ is measurable. Then if one of the following conditions is satisfied:

1) X^* is separable;

2) for any $\omega\in\Omega, F(\Omega)\in K_{kc}(X)$.

We obtain that F is a set-valued random variable.

From Lemma 3 and Lemma 5, when $X=R^d$, taking $F(\omega)\in K_{kc}(R^d)$, then for any $x^*\in R^d$, $F(\cdot)$ is measurable if and only if $S(x^*,F(\cdot))$ is $\mathcal{F}_t$ -measurable.

Lemma 6. (cf. [19]) Let $(\Omega,\mathcal{F})$ be a measurable space, X a separable metrizable space, and Y a metrizable space. Then every Caratheodory function $f:\Omega\times X\to Y$ (i.e. for each $x\in X$, the function $f(\cdot,x):\Omega\to Y$ is $\mathcal{F}$-measurable and for each $\omega\in\Omega$, the function $f(\omega,\cdot):X\to Y$ is continuous) is $(\mathcal{F}\otimes\mathcal{B}(X))$-mea- surable.

Theorem 2. Let $G\in M^1(\Omega\times[0,T],\Sigma,\nu_A;F_{kc}(R^d))$. Then for each $t\in[0,T]$, the fuzzy stochastic integral $\int_0^t G(s,\omega)\mathrm{d}A_s(\omega)$ is $\mathcal{F}_t$-measurable. Furthermore, the mapping $\psi(t,\omega)=\int_0^t G(s,\omega)\mathrm{d}A_s(\omega)$ is $\mathcal{B}([0,T])\otimes\mathcal{F}$-measurable.

Proof. Taking $G\in M^1(\Omega\times[0,T],\Sigma,\nu_A;F_{kc}(R^d))$, then for each $\alpha\in[0,1]$, the mapping $[G(t,\omega)]^\alpha:\Omega\times[0,T]\to\mathcal{K}(R^d)$ is Σ-measurable. For any $x^*\in R^d$, by Lemma 3, the support function $s(x^*,[G(t,\omega)]^\alpha)$ is Σ-measurable too. By Lemma 4, we have $s(x^*,\int_0^t[G(s,\omega)]^\alpha\,\mathrm{d}A_s)=\int_0^t s(x^*,[G(s,\omega)]^\alpha)\mathrm{d}A_s$. Since the real-valued Lebesgue-Stieltjes integral $\int_0^t s(x^*,[G(s,\omega)]^\alpha)\mathrm{d}A_s$ is a Carathedory function, then by Lemma 6, we obtain that $s(x^*,\int_0^t[G(s,\omega)]^\alpha\,\mathrm{d}A_s)$ is Σ-measurable. Therefore, by Lemma 5, for each $\alpha\in[0,1]$, the mapping $\int_0^t[G(s,\omega)]^\alpha\,\mathrm{d}A_s(\omega):\Omega\times[0,T]\to\mathcal{K}_{kc}(R^d)$ is $\mathcal{B}([0,T])\otimes\mathcal{F}$-measurable and $\mathcal{F}_t$-adapted, which means the integral $\psi(t,\omega)=\int_0^t G(s,\omega)\mathrm{d}A_s(\omega)$ is $\mathcal{B}([0,T])\otimes\mathcal{F}$-measurable and $\mathcal{F}_t$-adapted.

Theorem 3. Let $G\in M^2(\Omega\times[0,T],F_{kc}(R^d))$. Then for any $t\in[0,T]$, $\int_0^t G(s,\omega)\mathrm{d}A_s(\omega)\in L^2(\Omega,F(R^d))$.

Proof. By Theorem 2, for any $t\in[0,T]$, $\int_0^t G(s,\omega)\mathrm{d}A_s(\omega)$ is $\mathcal{F}_t$-measurable. We will show that for any $\omega\in\Omega$, $t\in[0,T]$, $E\left[\left\|\int_0^t[G(s,\omega)]^0\,\mathrm{d}A_s(\omega)\right\|_K^2\right]<\infty$.

For any $g\in S^2_{[G]^0}$,

$$\left\|\int_0^t g(s,\omega)\mathrm{d}A_s(\omega)\right\|^2 \le \mu_A([0,T])\int_0^t \|g_s(\omega)\|^2 \mathrm{d}A_s(\omega)$$
$$\le \mu_A([0,T])\int_0^t \left\|[G(s,\omega)]^0\right\|_K^2 \mathrm{d}A_s(\omega) \quad . (4)$$

Then

$$\sup_{g\in S_{[G]^0}} \left\|\int_0^t g_s(\omega)\mathrm{d}A_s(\omega)\right\|^2 \le \mu_A([0,T])\int_0^t \left\|[G(s,\omega)]^0\right\|_K^2 \mathrm{d}A_s.$$

Hence,

$$E\left[\left\|\int_0^t [G(s,\omega)]^0 \mathrm{d}A_s\right\|_K^2\right] \le E\left[\mu_A([0,T])\int_0^t \left\|[G(s,\omega)]^0\right\|_K^2 \mathrm{d}A_s\right]$$
$$\le \int_{\Omega\times[0,T]} \left\|[G(s,\omega)]^0\right\|_K^2 \mathrm{d}v_A < \infty, \quad (5)$$

which means $\int_0^t G(s,\omega)\mathrm{d}A_s(\omega) \in L^2(\Omega, F(R^d))$.

Theorem 4. Let $G \in M^1(\Omega\times[0,T], F_{kc}(R^d))$. Then for any $\omega\in\Omega$, $\int_0^t G(s,\omega)\mathrm{d}A_s(\omega)$ is continuous with respect to t under the metric d_∞.

Proof. Let $0 \le r < t \le T$, for any $\omega\in\Omega$, we have

$$\int_0^t G(s,\omega)\mathrm{d}A_s(\omega) = \int_0^r G(s,\omega)\mathrm{d}A_s(\omega) + \int_r^t G(s,\omega)\mathrm{d}A_s(\omega).$$

Then

$$d_\infty\left(\int_0^r G(s,\omega)\mathrm{d}A_s(\omega), \int_0^t G(s,\omega)\mathrm{d}A_s(\omega)\right)$$
$$= d_\infty\left(\int_0^r G(s,\omega)\mathrm{d}A_s(\omega), \int_0^r G(s,\omega)\mathrm{d}A_s(\omega) + \int_r^t G(s,\omega)\mathrm{d}A_s(\omega)\right)$$
$$\le d_\infty\left(\int_0^r G(s,\omega)\mathrm{d}A_s(\omega), \int_0^r G(s,\omega)\mathrm{d}A_s(\omega)\right) + d_\infty\left(\hat{\theta}, \int_r^t G(s,\omega)\mathrm{d}A_s(\omega)\right)$$
$$= d_\infty\left(\hat{\theta}, \int_r^t G(s,\omega)\mathrm{d}A_s(\omega)\right) = \sup_{\alpha\in[0,1]} d_H\left(\{0\}, \int_r^t [G(s,\omega)]^\alpha \mathrm{d}A_s(\omega)\right)$$
$$\le \sup_{\alpha\in[0,1]} \int_r^t d_H\left(\{0\}, [G(s,\omega)]^\alpha\right)\mathrm{d}A_s(\omega) \le \int_r^t \sup_{\alpha\in[0,1]} d_H\left(\{0\}, [G(s,\omega)]^\alpha\right)\mathrm{d}A_s(\omega)$$
$$\le \int_r^t \sup_{\alpha\in[0,1]} \left\|[G(s,\omega)]^\alpha\right\|_K \mathrm{d}A_s(\omega) = \int_r^t \left\|[G(s,\omega)]^0\right\|_K \mathrm{d}A_s(\omega) < \infty. \quad (6)$$

For any $\omega \in \Omega$, we have

$$\lim_{r \to t} \int_r^t \left\| \left[G(s,\omega) \right]^0 \right\|_K \mathrm{d}A_s(\omega) = 0.$$

Then for all $\omega \in \Omega$, $\int_0^t G(s,\omega)\mathrm{d}A_s(\omega)$ is left continuous for $t \in [0,T]$ under the metric d_∞. Similarly, we can prove that $\int_0^t G(s,\omega)\mathrm{d}A_s(\omega)$ is a right continuous for $t \in [0,T]$. Therefore it is continuous in t with respect to d_∞.

Lemma 7. Let fuzzy stochastic process $G \in L^1(\Omega \times [0,T], F_{kc}(R^d))$. Then for each $\alpha \in [0,1]$, there exists a sequence $\{f_\alpha^i : i \in N\} \subset S^1_{[G]^\alpha}$, such that for every $t \in [0,T]$,

$$S^1_{[G]^\alpha} = cl\{f_\alpha^i : i \in N\},$$

where the closure is taken in $L^1(\Omega \times [0,T], R^d; \nu_A)$.

Proof. Since $\mathcal{F}$ is separable with respect to probability measurable P, we have that $\mathcal{B}([0,T]) \otimes \mathcal{F}$ is separable with respect to product measure $\lambda \times P$. By Lemma 2, $L^1(\Omega \times [0,T], R^d; \lambda \times P)$ is separable. It can be obtained that $L^1(\Omega \times [0,T], \Sigma, R^d; \nu_A)$ is separable under the norm ν_A. So that for any $\alpha \in [0,1]$, $S^1_{[G]^\alpha}$ is separable since it is a closed subset of $L^1(\Omega \times [0,T], \Sigma, R^d; \nu_A)$. Then there exists a sequence $\{f_\alpha^i : i \in N\} \subset S^1_{[G]^\alpha}$,

$$S^1_{[G]^\alpha} = cl\{f_\alpha^i : i \in N\}$$

Theorem 5. For a fuzzy set-valued stochastic process $G \in M^1(\Omega \times [0,T], F_{kc}(R^d))$ and any $\alpha \in [0,1]$, there exists a sequence $\{f_\alpha^i : i \in N\} \subset S^1_{[G]^\alpha}$ such that

$$[G(t,\omega)]^\alpha = cl\{f_\alpha^i(t,\omega) : i \in N\} \ a.e. (t,\omega)$$

and for each t

$$\int_0^t [G(s,\omega)]^\alpha \, dA_s(\omega) = cl\left\{\int_0^t f_\alpha^i(s,\omega) dA_s(\omega) : i \in N\right\} \quad a.s.$$

where "cl" denotes the closure in R^d.

Proof. For each $\alpha \in [0,1]$, by Lemma 7, there exists a sequence $\{f_\alpha^n : n \in N\} \subset L^1(\Omega \times [0,T], \Sigma, R^d; \nu_A)$ such that

$$S^1_{[G]^\alpha} = cl\{f_\alpha^i : i \in N\},$$

where the closure is taken in $L^1(\Omega \times [0,T], \Sigma, R^d; \nu_A)$.

For each $\alpha \in [0,1]$, by Castaing represent theorem (cf. [15] [20]), there exists a sequence $\{g_\alpha^j : j \in N\} \subset S^1_{[G]^\alpha}$ such that

$$[G(t,\omega)]^\alpha = cl\{g_\alpha^j(t,\omega) : j \in N\} \text{ for all } (t,\omega) \in [0,T] \times \Omega.$$

At first we will show that

$$[G(t,\omega)]^\alpha \subset cl\{f_\alpha^i(t,\omega) : i \in N\} \; a.e. (t,\omega).$$

In fact, taking $g_\alpha^j \in S^1_{[G]^\alpha}$, there exists a sequence $\{f_\alpha^{i_k}\}$ such that

$$\left\| f_\alpha^{i_k} - g_\alpha^j \right\|_{L^1(\Omega \times [0,T], \Sigma, \nu_A, R^d)} \to 0,$$

then there exists a subsequence $\{f_\alpha^{i_{k_j}}\}$ such that

$$\left\| f_\alpha^{i_{k_j}}(t,\omega) - g_\alpha^j(t,\omega) \right\|_{R^d} \to 0 \; a.e (t,\omega).$$

Therefore

$$\left[G(t,\omega)\right]^{\alpha} \subset cl\left\{f_{\alpha}^{i}(t,\omega): i \in N\right\} \ a.e(t,\omega).$$

On the other hand

$$cl\left\{f_{\alpha}^{i}(t,\omega): i \in N\right\} \subset \left[G(t,\omega)\right]^{\alpha} \ a.e.(t,\omega),$$

since $\left[G(t,\omega)\right]^{\alpha}$ is closed and $f_{\alpha}^{i} \in S_{[G]^{\alpha}}^{1}$, which yields

$$\left[G(t,\omega)\right]^{\alpha} = cl\left\{f_{\alpha}^{i}(t,\omega): i \in N\right\} \ a.e(t,\omega).$$

Since

$$\int_0^t \left[G(s,\omega)\right]^{\alpha} \mathrm{d}A_s(\omega) = \left\{\int_0^t g_{\alpha}(s,\omega)\mathrm{d}A_s(\omega): g_{\alpha} \in S_{[G]^{\alpha}}^{1}\right\}$$

is closed and $f_{\alpha}^{i} \in S_{[G]^{\alpha}}^{1}$, then for each t

$$cl\left\{\int_0^t f_{\alpha}^{i}(s,\omega)\mathrm{d}A_s(\omega): i \in N\right\} \subset \int_0^t \left[G(s,\omega)\right]^{\alpha} \mathrm{d}A_s(\omega) \ a.s. \tag{7}$$

For any $g_{\alpha} \in S_{[G]^{\alpha}}^{1}$, there exists a sequence $\left\{f_{\alpha}^{i_k}: k \in N\right\} \subset \left\{f^{i}: i \in N\right\}$ such that

$$\left\|f_{\alpha}^{i_k} - g_{\alpha}\right\|_{L^1\left(\Omega\times[0,T],\Sigma,\nu_A,R^d\right)} \to 0.$$

Then for each t,

$$\int_0^t g_{\alpha}(s,\omega)\mathrm{d}A_s \in cl\left\{\int_0^t f_{\alpha}^{i}(s,\omega)\mathrm{d}A_s : i \in N\right\} a.s.$$

which means

$$\int_0^t \left[G(s,\omega)\right]^{\alpha} \mathrm{d}A_s(\omega) \subset cl\left\{\int_0^t f_{\alpha}^{i}(s,\omega)\mathrm{d}A_s(\omega): i \in N\right\} a.s. \tag{8}$$

(7) together with (8) yields

$$cl\left\{\int_0^t f_\alpha^i(s,\omega)\mathrm{d}A_s(\omega): i \in N\right\} = \int_0^t \left[G(s,\omega)\right]^\alpha \mathrm{d}A_s(\omega) \ a.s.$$

Lemma 8. (cf. [15]) Let $F \in M(\Omega, \mathcal{K}(R^d))$, $\phi: \Omega \times X \to \bar{R} = [-\infty, +\infty]$ satisfy: for fixed $\omega \in \Omega$, $\phi(\omega,\cdot)$ is continuous with respect to x, for fixed $x \in X$, $\phi(\omega,x)$ is measurable with respect to ω, then there exists an $f_0 \in S_F^p$ such that $\int_\Omega \phi(\omega, f(\omega))\mathrm{d}\mu < \infty$, then we have

$$\inf_{f \in S_F^p} \int_\Omega \phi(\omega, f(\omega))\mathrm{d}\mu = \int_\Omega \inf_{f \in F(\omega)} \phi(\omega, f(\omega))\mathrm{d}\mu.$$

Theorem 6. Let $F, G \in M^1\left(\Omega \times [0,T], F_{kc}(R^d)\right)$. Then for any $t \in [0,T]$,

$$d_\infty\left(\int_0^t F(s,\omega)\mathrm{d}A_s(\omega), \int_0^t G(s,\omega)\mathrm{d}A_s(\omega)\right) \le \int_0^t d_\infty\left(F(s,\omega), G(s,\omega)\right)\mathrm{d}A_s(\omega) \ a.s.$$

Proof. Let $\phi_t(\omega) = \int_0^t F_s(\omega)\mathrm{d}A_s(\omega), \psi_t(\omega) = \int_0^t G_s(\omega)\mathrm{d}A_s(\omega)$. By Theorem 5, we can obtain that for each $\alpha \in [0,1]$, there exist sequences $\{f_\alpha^i : i \ge 1\} \subseteq S^1_{[F]^\alpha}$ and $\{g_\alpha^j : j \ge 1\} \subseteq S^1_{[G]^\alpha}$ such that $[F(t,\omega)]^\alpha = cl\{f_\alpha^i(t,\omega) : i \ge 1\}\ a.e.(t,\omega)$, $[G(s,\omega)]^\alpha = cl\{g_\alpha^j(t,\omega) : j \ge 1\}\ a.e.(t,\omega)$. For each t,

$$\int_0^t \left[F(s,\omega)\right]^\alpha \mathrm{d}A_s(\omega) = cl\left\{\int_0^t f_\alpha^i(s,\omega)\mathrm{d}A_s(\omega): i \in N\right\} a.s.$$

and

$$\int_0^t \left[G(s,\omega)\right]^\alpha \mathrm{d}A_s(\omega) = cl\left\{\int_0^t g_\alpha^j(s,\omega)\mathrm{d}A_s(\omega): j \in N\right\} a.s.$$

Therefore

$$\begin{aligned}
&\inf_{y \in \int_0^t [G(s,\omega)]^\alpha \mathrm{d}A_s(\omega)} \left\| \int_0^t f_\alpha^i(s,\omega)\mathrm{d}A_s(\omega) - y \right\| \\
&= \inf_{j \ge 1} \left\| \int_0^t f_\alpha^i(s,\omega)\mathrm{d}A_s(\omega) - \int_0^t g_\alpha^j(s,\omega)\mathrm{d}A_s(\omega) \right\| \\
&\le \inf_{j \ge 1} \int_0^t \left\| f_\alpha^i(s,\omega) - g_\alpha^j(s,\omega) \right\| \mathrm{d}A_s(\omega) \ a.s.
\end{aligned} \tag{9}$$

By Lemma 8, we have

$$\inf_{j\geq 1}\int_0^t \left\| f_\alpha^i(s,\omega)-g_\alpha^j(s,\omega)\right\| \mathrm{d}A_s(\omega)=\int_0^t \inf_{j\geq 1}\left\| f_\alpha^i(s,\omega)-g_\alpha^j(s,\omega)\right\| \mathrm{d}A_s(\omega)\ a.s. \tag{10}$$

Then

$$\begin{aligned}
&\int_0^t \inf_{j\geq 1}\left\| f_\alpha^i(s,\omega)-g_\alpha^j(s,\omega)\right\| \mathrm{d}A_s(\omega)\\
&\leq \int_0^t \sup_{i\geq 1}\inf_{j\geq 1}\left\| f_\alpha^i(s,\omega)-g_\alpha^j(s,\omega)\right\| \mathrm{d}A_s(\omega)\ a.s.\\
&\leq \int_0^t d_H\left([F(s,\omega)]^\alpha,[G(s,\omega)]^\alpha\right)\mathrm{d}A_s(\omega)\ a.s.\\
&\leq \int_0^t \sup_{\alpha\in[0,1]} d_H\left([F(s,\omega)]^\alpha,[G(s,\omega)]^\alpha\right)\mathrm{d}A_s(\omega)\ a.s.\\
&=\int_0^t d_\infty\left(F(s,\omega),G(s,\omega)\right)\mathrm{d}A_s(\omega).
\end{aligned} \tag{11}$$

Then

$$\sup_{i\geq 1}\inf_{j\geq 1}\left\|\int_0^t f_\alpha^i(s,\omega)\mathrm{d}A_s(\omega)-\int_0^t g_\alpha^j(s,\omega)\mathrm{d}A_s(\omega)\right\|\leq \int_0^t d_\infty\left(F(s,\omega),G(s,\omega)\right)\mathrm{d}A_s(\omega)\ a.s.$$

Similarly, we have

$$\sup_{j\geq 1}\inf_{i\geq 1}\left\|\int_0^t g_\alpha^j(s,\omega)\mathrm{d}A_s(\omega)-\int_0^t f_\alpha^i(s,\omega)\mathrm{d}A_s(\omega)\right\|\leq \int_0^t d_\infty\left(F(s,\omega),G(s,\omega)\right)\mathrm{d}A_s(\omega)\ a.s.$$

Then for each $\alpha\in[0,1]$,

$$d_H\left(\int_0^t [F(s,\omega)]^\alpha \mathrm{d}A_s(\omega),\int_0^t [G(s,\omega)]^\alpha \mathrm{d}A_s(\omega)\right)\leq \int_0^t d_\infty\left(F(s,\omega),G(s,\omega)\right)\mathrm{d}A_s(\omega)\ a.s.$$

Therefore

$$\sup_{\alpha\in[0,1]} d_H\left(\int_0^t [F(s,\omega)]^\alpha \mathrm{d}A_s(\omega),\int_0^t [G(s,\omega)]^\alpha \mathrm{d}A_s(\omega)\right)\leq \int_0^t d_\infty\left(F(s,\omega),G(s,\omega)\right)\mathrm{d}A_s(\omega)\ a.s.$$

Hence

$$d_\infty\left(\int_0^t F(s,\omega)\mathrm{d}A_s(\omega), \int_0^t G(s,\omega)\mathrm{d}A_s(\omega)\right) \le \int_0^t d_\infty\left(F(s,\omega), G(s,\omega)\right)\mathrm{d}A_s(\omega)\ a.s.$$

Theorem 7. Let $F,G \in M^2\left(\Omega\times[0,T], F_{kc}\left(R^d\right)\right)$. Then for each $t\in[0,T]$, we have

$$d_\infty^2\left(\int_0^t F(s,\omega)\mathrm{d}A_s(\omega), \int_0^t G(s,\omega)\mathrm{d}A_s(\omega)\right) \le \mu_A([0,T])\int_0^t d_\infty^2\left(F(s,\omega), G(s,\omega)\right)\mathrm{d}A_s(\omega)\ a.s.$$

Proof. For any $\alpha\in[0,1]$, we have

$$\begin{aligned}
&\inf_{j\ge 1}\left\|\int_0^t f_\alpha^i(s,\omega)\mathrm{d}A_s(\omega) - \int_0^t g_\alpha^j(s,\omega)\mathrm{d}A_s(\omega)\right\|^2\\
&\le \inf_{j\ge 1}\left\{\int_0^t \left\|f_\alpha^i(s,\omega) - g_\alpha^j(s,\omega)\right\|\mathrm{d}A_s(\omega)\right\}^2\\
&\le \inf_{j\ge 1}\left\{\mu_A([0,T])\int_0^t \left\|f_\alpha^i(s,\omega) - g_\alpha^j(s,\omega)\right\|^2\mathrm{d}A_s(\omega)\right\}\ a.s.
\end{aligned} \tag{12}$$

by Lemma 8, we have

$$\begin{aligned}
&\inf_{j\ge 1}\left\{\mu_A([0,T])\int_0^t \left\|f_\alpha^i(s,\omega) - g_\alpha^j(s,\omega)\right\|^2\mathrm{d}A_s(\omega)\right\}\\
&= \mu_A([0,T])\inf_{j\ge 1}\int_0^t \left\|f_\alpha^i(s,\omega) - g_\alpha^j(s,\omega)\right\|^2\mathrm{d}A_s(\omega)\ a.s.\\
&= \mu_A([0,T])\int_0^t \inf_{j\ge 1}\left\|f_\alpha^i(s,\omega) - g_\alpha^j(s,\omega)\right\|^2\mathrm{d}A_s(\omega)\ a.s.
\end{aligned} \tag{13}$$

Then

$$\begin{aligned}
&\mu_A([0,T])\int_0^t \inf_{j\ge 1}\left\|f_\alpha^i(s,\omega) - g_\alpha^j(s,\omega)\right\|^2\mathrm{d}A_s(\omega)\\
&\le \mu_A([0,T])\int_0^t \sup_{i\ge 1}\inf_{j\ge 1}\left\|f_\alpha^i(s,\omega) - g_\alpha^j(s,\omega)\right\|^2\mathrm{d}A_s(\omega)\\
&\le \mu_A([0,T])\int_0^t d_H^2\left([F(s,\omega)]^\alpha, [G(s,\omega)]^\alpha\right)\mathrm{d}A_s(\omega)\\
&\le \mu_A([0,T])\int_0^t d_\infty^2\left(F(s,\omega), G(s,\omega)\right)\mathrm{d}A_s(\omega)\ a.s.
\end{aligned} \tag{14}$$

Then

$$\sup_{i\ge 1}\inf_{j\ge 1}\left\|\int_0^t f_\alpha^i(s,\omega)\mathrm{d}A_s(\omega) - \int_0^t g_\alpha^j(s,\omega)\mathrm{d}A_s(\omega)\right\|^2 \le \mu_A([0,T])\int_0^t d_\infty^2\left(F(s,\omega), G(s,\omega)\right)\mathrm{d}A_s(\omega)\ a.s.$$

Similarly, we have

$$\sup_{j\geq 1}\inf_{i\geq 1}\left\|\int_0^t g_\alpha^j(s,\omega)\mathrm{d}A_s(\omega)-\int_0^t f_\alpha^i(s,\omega)\mathrm{d}A_s(\omega)\right\|^2 \leq \mu_A([0,T])\int_0^t d_\infty^2(F(s,\omega),G(s,\omega))\mathrm{d}A_s(\omega)\ a.s.$$

Then for each $\alpha \in [0,1]$

$$d_H^2\left(\int_0^t [F(s,\omega)]^\alpha \mathrm{d}A_s(\omega),\int_0^t [G(s,\omega)]^\alpha \mathrm{d}A_s(\omega)\right)\leq \mu_A([0,T])\int_0^t d_\infty^2(F(s,\omega),G(s,\omega))\mathrm{d}A_s(\omega)\ a.s.$$

Moreover

$$\sup_{\alpha\in[0,1]} d_H^2\left(\int_0^t [F(s,\omega)]^\alpha \mathrm{d}A_s(\omega),\int_0^t [G(s,\omega)]^\alpha \mathrm{d}A_s(\omega)\right)\leq \mu_A([0,T])\int_0^t d_\infty^2(F(s,\omega),G(s,\omega))\mathrm{d}A_s(\omega)\ a.s.$$

Hence

$$d_\infty^2\left(\int_0^t F(s,\omega)\mathrm{d}A_s(\omega),\int_0^t G(s,\omega)\mathrm{d}A_s(\omega)\right)\leq \mu_A([0,T])\int_0^t d_\infty^2(F(s,\omega),G(s,\omega))\mathrm{d}A_s(\omega)\ a.s.$$

Remark 1. In Theorem 6 and Theorem 7, the inequalities hold too if we take the expectation on both sides.

4. CONCLUSION

In [21], the author studied the Lebesgue-Stieltjes integral of real stochastic processes with respect to fuzzy valued stochastic processes. In some references such as [5] [6] , the integrals of fuzzy stochastic processes with respect to time t and Brownian motion were studied. In order to guarantee measurability of the integral, Kim (2005) Li and Ren (2007) defined the integral indirectly by taking the decomposable closure. Here, when the integrand taked value in compact and convex subsets of $F(R^d)$, we defined directly the integral of fuzzy stochastic process with respect to real-valued finite variation processes by using selection method, which is different from the above references. Then we proved the measurability (Theorem 2), which was key and guaranteed the reasonability of the definition. Attribute to the good property of finite variation of integrator, the integral was bounded as and L^2-bounded under the metric d_∞ (Theorem 3, Theorem 6 and Theorem 7). This property was much well than the integral with respect to Brownian motion since the latter was of infinite variation. Thanks to the boundedness of the integral, it was possible to do the further work such as exploring solutions of fuzzy stochastic differential equations.

ACKNOWLEDGEMENTS

We thank the editor and the referees for their comments. This work is partly supported by NSFC (No. 11371135).

REFERENCES

1. Puri, M.L. and Ralescu, D.A. (1986) Fuzzy Random Variables. Journal of Mathematical Analysis and Applications, 114, 409-422.

2. Kisielewicz, M. (1997) Set-Valued Stochastic Integrals and Stochastic Inclusions. Stochastic Analysis and Applications, 15, 780-800.

3. Kim, B.K. and Kim, J.H. (1999) Stochastic Integrals of Set-Valued Processes and Fuzzy Processes. Journal of Mathematical Analysis and Applications, 236, 480-502

4. Jung, E.J. and Kim, J.H. (2003) On Set-Valued Stochastic Integrals. Stochastic Analysis and Applications, 21, 401-418.

5. Kim, J.H. (2005) On Fuzzy Stochastic Differentials Equations. Journal of the Korean Mathematical Society, 42, 153-169.

6. Li, S. and Ren, A. (2007) Representation Theorems, Set-Valued and Fuzzy Set-Valued ITO Integral. Fuzzy Sets and Systems, 158, 949-962.

7. Malionwski, M.T. and Michta, M. (2011) On Set-Valued Stochastic Integrals and Fuzzy Stochastic Equations. Fuzzy Sets Systems, 177, 1-19.

8. Zhang, J., Li, S., Mitoma, I. and Okazaki, Y. (2009) On Set-Valued Stochastic Integrals in an M-Type 2 Banach Space. Journal of Mathematical Analysis and Applications, 350, 216-233.

9. Zhang, J., Li, S., Mitoma, I. and Okazaki, Y. (2009) On the Solution of Set-Valued Stochastic Differential Equations in M-Type 2 Banach Space. Tohoku Mathematical Journal, 61, 417-440.

10. Li, J. and Wang, J. (2012) Fuzzy Set-Valued Stochastic Lebesgue Integral. Fuzzy Sets and Systems, 200, 48-64.

11. Mitoma, I., Okazaki, Y. and Zhang, J. (2010) Set-Valued Stochastic Differential Equations in M-Type 2 Banach Space. Communications on Stochastic Analysis, 4, 215-237.

12. Zhang, J., Mitoma, I. and Okazaki, Y. (2013) Set-Valued Stochastic Integral with Respect to Poisson Process in a Banach Space. International Journal of Approximate Reasoning, 54, 404-417.

13. Fei, W. (2013) Existence and Uniquess for Solutions to Fuzzy Stochastic Differential Equations Driven by Local Martingales under the Non-Lipschtiz Condition. Nolinear Analysis, 76, 202-214.

14. Zhang, J. and Qi, J. (2013) Set-Valued Stochastic Integrals with Respect to Finite Variation Processes. Advances in Pure Mathematics, 3, 15-19.

15. Hu, S. and Papageorgiou, N. (1997) Handbook of Multivalued Analysis, Volume I: Theory. Kluwer Academic Publishers, Boston.

16. Negoito, C. and Ralescu, D. (1975) Applications of Fuzzy Sets to Systems Analysis. Wiley, New York.

17. Kwakernaak, H. (1987) Fuzzy Random Variables, Definition and Theoroems. Information Sciences, 15, 1-29.

18. Zhang, W., Li, S. and Wang, Z. (2007) Set-Valued Stochastic Process Introduction. Science Press, Beijing. (In Chinese)

19. Charalambos, D.A. and Kim, C.B. (1994) Infinite Dimensional Analysis. Springer-Verlag, Berlin.

20. Li, S., Ogura, Y. and Kreinovich, Y. (2002) Limit Theorems and Applications of Set-valued and Fuzzy Set-Valued Random Variables. 43rd Edition, Kluwer Academic Publishers, Dordrecht.

21. Feng, Y. (2001) Fuzzy-Valued Mappings with Finite Variation, Fuzzy-Valued Measures and Fuzzy-Valued Lebesgue-Stieltjes Integrals. Fuzzy Sets and Systems, 2, 227-236.

CHAPTER 5

Numerical Approximation of Fractal Dimension of Gaussian Stochastic Processes

Freddy H. Marin Sanchez[1]**, William Eduardo Alfonso**[2]

Basic Science Department, EAFIT University, Carrera 49 # 7 Sur-50, Medellín, Colombia

ABSTRACT

In this paper we propose a numerical method to estimate the fractal dimension of stationary Gaussian stochastic processes using the random Euler numerical scheme and based on an analytical formulation of the fractal dimension for filtered stochastic signals. The discretization of continuous time processes through this random scheme allows us to find, numerically, the expected value, variance and correlation functions at any point of time. This alternative method for estimating the fractal dimension is easy to implement and requires no sophisticated routines. We use simulated data sets for stationary processes of the type Random Ornstein Uhlenbeck to graphically illustrate the results and compare them with those obtained whit the box counting theorem.

KEYWORDS

Stationary Gaussian Stochastic Processes, Fractal Dimension, Random Euler Numerical Scheme

1. INTRODUCTION

A common practice when trying to measure irregular shapes, such as the perimeter of an island or the length of a coastline, is to use Euclidean geometry, ignoring the fact that these shapes do not correspond to the ones of ideal objects, such as polygons and circles, whose dimension is an integer. In contrast, many forms observed in nature have a special feature that makes their resulting measurement depend on the measuring scale, such that the lower the scale, the higher the value of the measurement, to the point where it becomes infinite or indeterminate. The dimension of irregular shapes is a non-integer dimension known as the Hausdorff-Besicovitch dimension. In the early twentieth century, Lewis Fry Richardson found a relationship that allows determining the value of a constant that indicates the degree of roughness of a coast or a geographical border, and used it to calculate the border length between several countries [1] . Based on Richardson's findings, Benoit Mandelbrot [2] found that such irregular shapes have the property of statistical self-similarity, which states that any portion of an irregular shape can be considered a fully scaled image of the entire shape. Later, Mandelbrot coined this property as the fractal dimension to describe sets in $\mathbb{R}^d$. There are many studies in the literature in which maps, border areas and natural and physical phenomena are modeled and simulated with fractal features. In [3] for example, a proposal to simulate real maps using the fractal dimension of Gaussian stochastic processes is presented and some reports modeling terrestrial surfaces, irregular oceanic coasts and rivers, can be found in [4] -[9] .

There are several approaches to estimating of the fractal dimension. In fact in [10] different types of fractals that appear in scientific research are examined, application in nonlinear dynamical systems is discussed, and an exahustive review of numerical methods for their estimation is done.

[11] proposed another method for estimating the fractal dimension of subset of $\mathbb{R}^d$, as well as an additional correction to address the problem of resolution, allowing obtaining not only the fractal dimension, but also a confidence interval for this estimation, thus improving the box counting method. They also show some experimental results based on simulated data and real data from the coastline of Norway. The techniques for the measurement of fractal dimension can be divided into two groups: those based on vectors, which include structured algorithms, and those based on matrices, for the analysis of image systems that include the distance transformation method reported in [12] . Moreover, [13] introduces the roughness method in order to determine the fractal dimension of the Sierpinski carpet, triangle and tetrahedron, the standard Cantor set and the Menger sponge. Meanwhile, in the study of stochastic signals filtering by [14] is obtained the analytical formulation of the fractal dimension to investigate the effects of a linear filter and the regularity in a particular mono-fractal stochastic process, based on an explicit, continuous formulation, linking the spectral properties of a given filter and the fractal properties of a given process, knowing the function of probability density and output, departing from the correlation coefficient of the filter. With regard to the arguments established by [14] , the aim of our study is to estimate, numerically, the fractal dimension of ergodic stationary Gaussian stochastic processes using the random Euler scheme.

This paper is organized as follows. Section 2 presents the Euler random scheme and builds the numerical solution, the expected value and the crossed expected value for a more general stochastic processe that allows calculating the variance and correlation of the solution of problems of random initial value; the definition of fractal dimension and the box counting theorem are described in Section 3. In Section 4 we present a numericalanalytical formulation for calculating length and fractal dimension for stationary Gaussian stochastic processes. A method to calculate the length and fractal dimension of stationary Gaussian stochastic processes that include processes such as random Ornstein Uhlenbeck is presented in Section 5. Section 6 illustrates graphically, some experimental numerical results associated with the calculation of the fractal dimension of ergodic stationary Gaussian stochastic processes, compared with the results obtained from using the box count theorem.

2. ON THE RANDOM EULER NUMERICAL SCHEME

Consider the initial random value problem of the form[1],

$$\dot{X}(t) = f(X(t),t), \quad t \in T = [t_0, t_1]$$
$$X(t_0) = X_0 \tag{1}$$

where X_0 is a random variable and therefore $X(t)$ as $f(X(t),t)$ are both unknown stochastic processes defined on the same probability space $(\Omega, \mathcal{F}, \mathbb{P})$. The aim of this section is to use the Euler scheme to construct a numerical solution, the expected value and the crossed expected value to random initial value problem (1).

Note that the discretized solution of Equation (1) can be expressed in integral form as

$$X_{t_n} - X_{t_{n-1}} = \int_{t_{n-1}}^{t_n} f(X(s), S)\,ds.$$

By using the approximation $\int_{t_{n-1}}^{t_n} f(S, X(s))ds \approx f(t_{n-1}, X(t_{n-1}))(t_n - t_{n-1})$, the numerical scheme

$$\begin{cases} X_{t_n} = X_{t_{n-1}} + hf(X_{t_{n-1}}, t_{n-1}) \\ X_{t_0} = X_0 \end{cases} \quad n \geq 1, \tag{2}$$

is obtained, where X_{t_n}, $f(X_{t_{n-1}}, t_{n-1})$ are second order random variables (2-r.v.) and $h = t_n - t_{n-1}$.

In particular, consider a (2-r.v.) X_0 and a stochastic process $b(t)$ second order (2-s.p.) m.s. integrable defined on the same probability space $(\Omega, \mathcal{F}, \mathbb{P})$ such that $b(t)$ is independent of X_0 for each t lying in the interval $T = [t_0, t_1]$. Let $a(t)$ a continuous deterministic function defined on T. The idea is to solve numerically initial random value problems given by the expression

$$\dot{X}(t) = a(t)X(t) + b(t), \quad t \in T = [t_0, t_1]$$
$$X(t_0) = X_0. \tag{3}$$

Note that the Euler numerical scheme for problem (3)[2] takes the form

$$\begin{aligned} X_{t_n} &= X_{t_{n-1}} + h\left[a(t_{n-1})X_{t_{n-1}} + b(t_{n-1})\right], \\ &= \left(1 + ha(t_{n-1})\right)X_{t_{n-1}} + hb(t_{n-1}) \quad n \geq 1. \end{aligned} \tag{4}$$

It is easy to show using the principle of mathematical induction, (see [19] and [20] pp. 58-59), that X_{t_n} and $X_{t_n}X_{t_m}$ can be expressed as

$$X_{t_n} = X_{t_0}\prod_{i=0}^{n-1} f_i + h\sum_{i=0}^{n-1}\left(\prod_{j=i+1}^{n-1} f_j\right)b(t_i); \qquad n \geq 1. \tag{5}$$

$$\begin{aligned} X_{t_n}X_{t_m} &= X_{t_0}^2\left(\prod_{i=0}^{n-1} f_i\right)\left(\prod_{j=0}^{m-1} f_j\right) + h^2\sum_{i=0}^{n-1}\sum_{k=0}^{m-1}\left(\prod_{j=i+1}^{n-1} f_j\right)\left(\prod_{l=k+1}^{m-1} f_l\right)b(t_i)b(t_k) \\ &+ hX_{t_0}\left[\left(\prod_{i=0}^{n-1} f_i\right)\sum_{k=0}^{m-1}\left(\prod_{l=k+1}^{m-1} f_l\right)b(t_k) + \left(\prod_{k=0}^{m-1} f_k\right)\sum_{i=0}^{n-1}\left(\prod_{j=k+1}^{n-1} f_j\right)b(t_i)\right], \end{aligned} \tag{6}$$

where $f_i = 1 + ha(t_i)$. From Equations (5), (6) and the results reported in [19] , it follows that the expected values $E\{X_{t_n}\}$ and $E\{X_{t_n}X_{t_m}\}$ takes the form

$$E\{X_{t_n}\} = E\{X_{t_0}\}\left(\prod_{i=0}^{n-1} f_i\right) + h\sum_{i=0}^{n-1}\left(\prod_{j=i+1}^{n-1} f_j\right)E\{b(t_i)\}, \tag{7}$$

$$\begin{aligned} E\{X_{t_n}X_{t_m}\} &= E\{X_{t_0}^2\}\left(\prod_{i=0}^{n-1} f_i\right)\left(\prod_{i=0}^{n-1} f_k\right) + h^2\sum_{i=0}^{n-1}\sum_{k=0}^{m-1}\left(\prod_{j=i+1}^{n-1} f_j\right)\left(\prod_{l=k+1}^{m-1} f_l\right)E\{b(t_i)b(t_k)\} \\ &+ hE\{X_{t_0}\}\left[\left(\prod_{i=0}^{n-1} f_i\right)\sum_{k=0}^{m-1}\left(\prod_{l=k+1}^{m-1} f_l\right)E\{b(t_k)\} + \left(\prod_{k=0}^{m-1} f_k\right)\sum_{i=0}^{n-1}\left(\prod_{l=k+1}^{n-1} f_j\right)E\{b(t_i)\}\right]. \end{aligned} \tag{8}$$

3. FRACTAL DIMENSION AND THE BOX COUNTING THEOREM

The fractal dimension is an objective way to compare fractals geometrically. This section introduces the concept of fractal dimension and presents the box counting theorem.

Let (X,d) denote a complete metric space. Let $A \in H(X)$ be a nonempty compact subset of X. Let $B(x,\epsilon)$ denote the closed ball of radius $\epsilon > 0$ and center at a point $x \in X$. We wish to define an integer, $N(A,\epsilon)$, to be the least number of closed balls of radius ϵ needed to cover the set A.

That is, $N(A,\epsilon) = M$: smallest positive integer M such that $A \subset \bigcup_{n=1}^{M} B(x_n,\epsilon)$, for some set of distinct points $\{x_n : n = 1,2,\cdots,M\} \subset X$. In other words, surround every point $x \in A$ by an open ball of radius $\epsilon > 0$ to provide a cover of A by open sets. Because A is compact this cover possesses a finite subcover, consisting of an integer number, say $\hat{M}$, of open balls. By taking the closure of each ball, thus obtained a cover consisting of M closed balls. Let C denote the set of covers of A by at most $\hat{M}$ closed balls of radius ϵ. Then C contains at least one element.

Let $f : C \to \{1,2,3,\cdots,M\}$ defined by $f(c) =$ number of balls in the cover $c \in C$. Then $\{f(c) : c \in C\}$ is a finite set of positive integers. It follows that it contains a least integer, $N(A,\epsilon)$.

The intuitive idea behind fractal dimension is that a set A has fractal dimension D if: $N(A,\epsilon) \approx C\epsilon^{-D}$ for some positive constant C [3].

Accordingly,

$$D \approx \frac{\ln N(A,\epsilon) - \ln C}{\ln(1/\epsilon)}.$$

Now notice that the term $\frac{\ln C}{\ln(1/\epsilon)}$ approaches 0 as $\epsilon \to 0$. This leads us to the following definition.

Definition 3.1. Let $A \in H(X)$ where (X,d) is a metric space. For each $\epsilon > 0$, let $N(A,\epsilon)$ denote the smallest number of closed balls of radius $\epsilon > 0$ needed to cover A. If

$$D = \lim_{\epsilon \to 0} \left\{ \frac{\ln N(A,\epsilon)}{\ln(1/\epsilon)} \right\}$$

exists, then D is called the fractal dimension A. Also used the notation $D = D_A$ and will say " A has fractal dimension D".

Theorem 3.2. The Box Counting Theorem. Let $A \in H(\mathbb{R}^m)$, where the Euclidean metric is used. Cover $\mathbb{R}^m$ by closed square boxes of side length $(1/2^n)$. Let $N_n(A)$ denote the number of boxes of side length $(1/2^n)$ which intersect the attractor. If

$$D = \lim_{n \to \infty} \left\{ \frac{\ln(N_n(A))}{\ln(2^n)} \right\},$$

then A has fractal dimension D. The proof of this theorem and other important definitions can be found at [21] .

4. ANALYTICAL FORMULATION OF THE FRACTAL DIMENSION FOR STOCHASTIC PROCESSES

An analytical expression for the fractal dimension of filtered stochastic signals with the purpose of studying the effects of a linear filter and the regularity of a given stochastic mono-fractal process is introduced in [14]. They present some arguments using the Euclidean length of a deterministic curve $x(t)$ with limited duration T, as starting point to estimate the length and fractal dimension of such stochastic signals.

However, stochastic processes can not be described by deterministic continuous curves and irregularities are present even when examined with higher resolutions. This leads to the paradoxical conclusion that the lengths of such signals are not finite. To overcome the fact that stochastic processes lack of a finite longitude, [2] assumes that the longitude L_x of a fractal curve depends on the resolution and the length of the measurement instrument, λ. [1] proves that measuring $L_x(\lambda)$ as a function of λ allows to determine the fractal dimension i.e.:

$$L_x(\lambda) \approx K_x \lambda^{1-D_x} \tag{9}$$

where D_x is the fractal dimension of the curve and K_x is a constant.

Thus, the length L_x can be written as a function of λ given by

$$L_x(\lambda) \approx T\overline{\lambda(h)} \tag{10}$$

where $\overline{\lambda(h)} = \frac{1}{T}\int_{-T/2}^{T/2} \lambda(t,h)\,dt$ and where the elementary length is given by $\lambda(t,h) \approx (1/h)\sqrt{h^2 + \Delta(t,h)^2}$.

This definition of length for continuous curves can be extended to ergodic stationary stochastic processes by replacing the arithmetic mean by the mathematical expectation $E\{\cdot\}$, i.e.

$$L_x(\lambda) = TE[\lambda]$$

where $\lambda = (1/h)\sqrt{h^2 + \Delta^2}$.

Note that $L_x(h) = \frac{T}{h}E\left\{\sqrt{h^2 + \Delta^2}\right\}$ only is solved numerically. In particular, for a high frequency sampling it can be assumed that $h \ll \Delta$ to obtain an analytical expression given by [22]

$$L_x(h) \approx \frac{T}{h}E\{|\Delta|\}. \tag{11}$$

Length and Fractal Dimension for Gaussian Stochastic Processes

Following the results reported in [14] , we will make a brief extension that also requires a prior information such as the probability density function of the process and the variance and correlation functions.

Consider a stochastic process $\{X_t\}_{t\geq 0}$ with values in $\mathbb{R}$ defined on a probability space $(\Omega, \mathcal{F}, \mathbb{P})$. The average length of the process can be estimated by the expression

$$L_x(t) \approx \frac{T}{h} E\{|x_1(t) - x_2(t)|\} \tag{12}$$

with $x_1(t) = X_t, x_2(t) = X_{t-h}; t, t-h \in [0,T]$. Thus, the expected value of the absolute value of the difference between $x_1(t)$ and $x_2(t)$ takes the form

$$E\{|x_1(t) - x_2(t)|\} = \iint |x_1(t) - x_2(t)| \, p_{x_1,x_2}(x_1(t), x_2(t)) \, \mathrm{d}x_1 \mathrm{d}x_2, \tag{13}$$

where, $p_{x_1,x_2}(x_1(t), x_2(t))$ denotes the joint probability density function.

Now suppose that the stochastic process $\{X_t\}_{t\geq 0}$ is a stationary Gaussian process. Consequently, $x_1(t)$ and $x_2(t)$ are also Gaussian stationary and therefore the joint probability density can be written as

$$p_{x_1,x_2}(x_1(t), x_2(t)) = \frac{\left(1-\rho_x^2(t)\right)^{1/2}}{2\pi\sigma_1(t)\sigma_2(t)} \mathrm{e}^{\left(-1/\left(2\left(1-\rho_x^2(t)\right)\right)\right)\left(x_1^2(t)/\sigma_1^2(t) + x_2^2(t)/\sigma_2^2(t) - 2\rho_x(t)x_1(t)x_2(t)/\sigma_1(t)\sigma_2(t)\right)}, \tag{14}$$

where $\sigma_1^2(t)$ and $\sigma_2^2(t)$ are the variances of $x_1(t)$ and $x_2(t)$ respectively, and $\rho_x(t)$ is the correlation function defined by $\rho_x(t) = E\{x_1(t)x_2(t)\} / \sqrt{E\{x_1^2(t)\}E\{x_2^2(t)\}}$.

The probability density of the increment $\Delta = x_1(t) - x_2(t)$ is

$$p_\Delta(\Delta)=\int_{-\infty}^{+\infty}p_{x_1x_2}(x_1(t),x_1(t)+\Delta)\mathrm{d}x_1. \tag{15}$$

Analogously to [14] , the probability density of the increment follows a normal distribution $N(0,\tilde{\sigma}(t)^2)$:

$$p_\Delta(\Delta)=\frac{1}{\sqrt{2\pi\tilde{\sigma}(t)}}\mathrm{e}^{-\Delta^2/2\tilde{\sigma}(t)^2}, \tag{16}$$

with $\tilde{\sigma}(t)=\sqrt{\sigma_1^2(t)+\sigma_2^2(t)-2\rho_x(t)\sigma_1(t)\sigma_2(t)}$. As $x_1(t)$ and $x_2(t)$ belong to the same process $\{X_t\}_{t\geq 0}$ then $\sigma_1(t)=\sigma_2(t)=\sigma_x(t)$ and the standard deviation is $\tilde{\sigma}(t)=\sigma_x(t)\sqrt{2(1-\rho_x(t))}$.

The mathematical expectation of the increment becomes

$$E\{|\Delta|\}=2\int_0^{+\infty}\Delta p_\Delta(\Delta)\mathrm{d}\Delta=\sqrt{\frac{2}{\pi}}\tilde{\sigma}(t), \tag{17}$$

and thus

$$L_x(t)\approx\frac{2}{\sqrt{\pi}}\frac{T\sigma_x(t)}{h}\sqrt{1-\rho_x(t)};\quad \forall t\in[0,T]. \tag{18}$$

Using Equation (9), the fractal dimension D_x of the process $\{X_t\}_{t\geq 0}$ can be approximated as

$$D_x(t,h)\approx 1-\frac{\log(L_x(t))}{\log(h)}+\frac{\log(K)}{\log(h)}. \tag{19}$$

Finally, Equation (18) becomes

$$D_x(t,h)=2-\frac{1}{2}\lim_{h\to 0}\frac{\log(1-\rho_x(t))}{\log(h)}. \tag{20}$$

5. NUMERICAL APPROACH TO FRACTAL DIMENSION FOR STATIONARY GAUSSIAN PROCESSES

In this section we present the numerical construction of the fractal dimension for stationary Gaussian stochastic processes including random Ornstein Uhlenbeck process, using the results of Section 4, the random Euler numerical scheme and the approximation of correlation function using the calculations for $E\{X_{t_n}\}$ and $E\{X_{t_n}X_{t_m}\}$ described in Section 2.

5.1. Calculation of Fractal Dimension for the Random Ornstein Uhlenbeck Process

Consider the random initial value problem given by

$$\dot{X}(t) = -\alpha X(t) + \sigma W(t); \; t \in [0,T] \tag{21}$$

$$X(0) = X_0$$

where $\alpha > 0$, $\sigma > 0$, and $W(t)$ is a Gaussian white noise independent of X_0 for each t lying in the interval $[0,T]$ (defined as in [16] p. 121) and $X_0 \sim N\left(0, \frac{\sigma^2}{2\alpha}\right)$. Note that this process is stationary Gaussian and ergodic, and can be considered as a modification of the random walk in continuous time and analog to the order 1 autoregressive process, AR (1) with discrete time, that broadly describes the velocity of a Brownian mass particle under the influence of friction [23] .

Under random Euler numerical scheme (5), the Equation (21) can be expressed as

$$X_{t_n} = \prod_{i=0}^{n-1}(1-\alpha h)X_0 + \sigma h \sum_{i=0}^{n-1}\left(\prod_{j=i+1}^{n-1}(1-\alpha h)\right)W(t_i); \quad 1 \le n \le N; h = \frac{T}{N}. \tag{22}$$

$$X_{t_n}=\beta^n X_0+\sigma\sqrt{h}\sum_{i=0}^{n-1}\beta^{n-1-i}\xi_i, n\geq 1,\quad \xi_i\sim N(0,1),$$

where $\beta=1-\alpha h$. By Equation (8) it follows that

$$E\{X_{t_n}X_{t_m}\}=\beta^{n+m}E\{X_0^2\}+\sigma^2 hE\left\{\sum_{i=0}^{n-1}\beta^{n-1-i}\xi_i\sum_{j=0}^{m-1}\beta^{m-1-j}\xi_j\right\}.$$

Trivially note that $E\{X_{t_n}\}=0$. For $\beta<1$, $\sum_{i=0}^{n-1}\beta^i=\frac{1-\beta^n}{1-\beta}$; thus

$$E\{X_{t_n}X_{t_m}\}=\beta^{n+m}+\sigma^2 h\left(\frac{1-\beta^{2\min\{n,m\}}}{1-\beta^2}\right).$$

For the case in which $n=N$ and $m=N-1$ it follows that

$$E\{X_{t_N}X_{t_{N-1}}\}=\beta^{2N-1}+\sigma^2 h\left(\frac{1-\beta^{2(N-1)}}{1-\beta^2}\right)=\beta E\{X_{t_{N-1}}^2\}.$$

It is easy to see that $\lim_{N\to\infty}E\{X_{t_N}^2\}=\lim_{N\to\infty}E\{X_{t_{N-1}}^2\}=\frac{\sigma^2 h}{1-\beta^2}$, and $\lim E\{X_{t_N}X_{t_{N-1}}\}=\beta\frac{\sigma^2 h}{1-\beta^2}$. In this way,

$$\rho(h)=\frac{\beta\frac{\sigma^2 h}{1-\beta^2}}{\sqrt{\left(\frac{\sigma^2 h}{1-\beta^2}\right)\left(\frac{\sigma^2 h}{1-\beta^2}\right)}}=\beta,$$

which by Equation (20) becomes

$$D_x(h)=2-\frac{1}{2}\lim_{h\to 0}\frac{\log(\alpha h)}{\log(h)}. \tag{23}$$

5.2. Calculation of Fractal Dimension for a Random Mean Reverting Process

Consider the random initial value problem given by

$$\dot{X}(t) = \alpha(\mu - X(t)) + \sigma W(t); \; t \in [0,T] \tag{24}$$

$$X(0) = X_0,$$

where $\alpha > 0$, $\sigma > 0$, $\mu \in \mathbb{R}$, $W(t)$ is a Gaussian white noise independent of X_0 for each t lying in the interval $[0,T]$ (defined as in [16] p. 121) and $X_0 \sim N\left(0, \frac{\sigma^2}{2\alpha}\right)$ with $\sigma^2 > 2\alpha\mu^2$. Note that this process is stationary Gaussian and ergodic. When the initial condition is constant, this process is also known as mean reversion process with additive noise. This kind of process has been widely used in modeling the dynamic behavior of commodity prices and short term interest rates (see [24] and references therein), and the simulating of realistic irregular lattices in [3] .

Equation (24) can be written as

$$\dot{X}(t) = -\alpha X(t) + b(t); \quad b(t) = \alpha\mu + \sigma W(t). \tag{25}$$

Under random Euler numerical scheme, Equation (25) can be expressed as

$$X_{t_n} = \prod_{i=0}^{n-1}(1-\alpha h) X_0 + h\sum_{i=0}^{n-1}\left(\prod_{j=i+1}^{n-1}(1-\alpha h)\right)(\alpha\mu + \sigma W(t_i)); \quad 1 \le n \le N; h = \frac{T}{N}. \tag{26}$$

$$X_{t_n} = \theta\sum_{i=0}^{n-1}\beta^i + \beta^n X_0 + \sigma\sqrt{h}\sum_{i=0}^{n-1}\beta^{n-1-i}\xi_i, n \ge 1, \quad \xi_i \sim N(0,1).$$

Equation (8) implies that for $\theta = \alpha\mu h$ and $\beta = 1 - \alpha h$,

$$E\{X_{t_n} X_{t_m}\} = \theta^2 \sum_{i=0}^{n-1}\beta^i \sum_{j=0}^{m-1}\beta^j + \beta^{n+m} E\{X_0^2\} + \theta E\{X_0\}\left(\beta^m \sum_{i=0}^{n-1}\beta^i + \beta^n \sum_{j=0}^{m-1}\beta^j\right) + \sigma^2 h E\left\{\sum_{i=0}^{n-1}\beta^{n-1-i}\xi_i \sum_{j=0}^{m-1}\beta^{m-1-j}\xi_j\right\}.$$

Note that for $\beta<1$, $\sum_{i=0}^{n-1}\beta^i=\frac{1-\beta^n}{1-\beta}$; in this way, $E\{X_n\}=\theta\left(\frac{1-\beta^n}{1-\beta}\right)+\beta^n E\{X_0\}$.

$$E\{X_{t_n}X_{t_m}\}=\theta^2\left(\frac{1-\beta^n}{1-\beta}\right)\left(\frac{1-\beta^m}{1-\beta}\right)+\beta^{n+m}E\{X_0^2\}+\theta E\{X_0\}\left(\beta^m\left(\frac{1-\beta^n}{1-\beta}\right)+\beta^n\left(\frac{1-\beta^m}{1-\beta}\right)\right)+\sigma^2 h\left(\frac{1-\beta^{2\min\{n,m\}}}{1-\beta^2}\right)$$

For the case in which $n=N$ and $m=N-1$ it follows that $E\{X_{t_N}X_{t_{N-1}}\}=\beta E\{X_{t_{N-1}}^2\}+\theta E\{X_{t_{N-1}}\}$.

It is easy to see that $\lim_{N\to\infty}E\{X_{t_N}\}=\frac{\theta}{1-\beta}$; $\lim_{N\to\infty}E\{X_{t_N}^2\}=\lim_{N\to\infty}E\{X_{t_{N-1}}^2\}=\frac{\theta^2}{(1-\beta)^2}+\frac{\sigma^2 h}{1-\beta^2}$ and

$\lim_{N\to\infty}E\{X_{t_N}X_{t_{N-1}}\}=\beta\frac{\theta^2}{(1-\beta)^2}+\beta\frac{\sigma^2 h}{1-\beta^2}+\frac{\theta^2}{1-\beta}$. In this way,

$$\rho(h)=\frac{\beta\frac{\theta^2}{(1-\beta)^2}+\beta\frac{\sigma^2 h}{1-\beta^2}+\frac{\theta^2}{1-\beta}}{\sqrt{\left(\frac{\theta^2}{(1-\beta)^2}+\frac{\sigma^2 h}{1-\beta^2}\right)\left(\frac{\theta^2}{(1-\beta)^2}+\frac{\sigma^2 h}{1-\beta^2}\right)}}=\frac{\left(\mu^2+\frac{\sigma^2}{2\alpha-\alpha^2 h}\right)(1-\alpha h)+\alpha\mu^2 h}{\left(\mu^2+\frac{\sigma^2}{2\alpha-\alpha^2 h}\right)}=1-\frac{\alpha\sigma^2 h}{\sigma^2+\mu^2\left(2\alpha-\alpha^2 h\right)}$$

Equation (20) becomes

$$D_x(h)=2-\frac{1}{2}\lim_{h\to 0}\left[\frac{\log\left(\frac{\alpha\sigma^2 h}{\sigma^2+\mu^2\left(2\alpha-\alpha^2 h\right)}\right)}{\log(h)}\right]. \tag{27}$$

6. NUMERICAL RESULTS

This section graphically illustrates some experimental numerical results related to the calculation of the fractal dimension for stationary Gaussian stochastic processes associated with random initial value problems (21) and (24).

Figure 1(a) shows a comparison between the approximate average fractal dimension of 1000 simulated trajectories for Ornstein Uhlenbeck process using the Box Counting theorem and the numerical fractal dimension proposed in Equation (23) for different values of h. In Figure 1(b) the dynamic behavior of the fractal dimension is shown for different values of the parameter α .

Figure 2(a) shows a comparison between the approximate average fractal dimension of 1000 simulated trajectories for mean reverting process using the Box Counting theorem and the numerical fractal dimension proposed in Equation (27) for different values of h. In Figure 2(b) the dynamic behavior of the fractal dimension is shown for different values of the parameter α .

7. CONCLUSIONS

This paper made a slight extension of the approach proposed in [14] for the numerical estimation of the fractal dimension of ergodic stationary Gaussian stochastic processes associated with the random Ornstein Uhlenbeck process. Using random Euler scheme performed numerical construction of the expectation value, variance and correlation functions at any point of time to initial random value problems general. The computational implementation of this alternative method to estimate the fractal dimension is easy and requires no sophisticated routines. It illustrates graphically the results obtained compared with box counting theorem using simulated data sets for stationary processes type random Ornstein Uhlenbeck.

A clear convergence of the proposed methods is shown in Figure 1(a) and Figure 2(a). In addition, the results show that the fractal dimension of Ornstein Uhlenbeck processes is independent of the parameter μ . This independence is maintained in the mean reverting process for the case where $\sigma^2 > 2\alpha\mu^2$. Figure 1(b) and Figure 2(b) reflect the fact that the fractal dimension of this type of process is more sensitive to the parameter α .

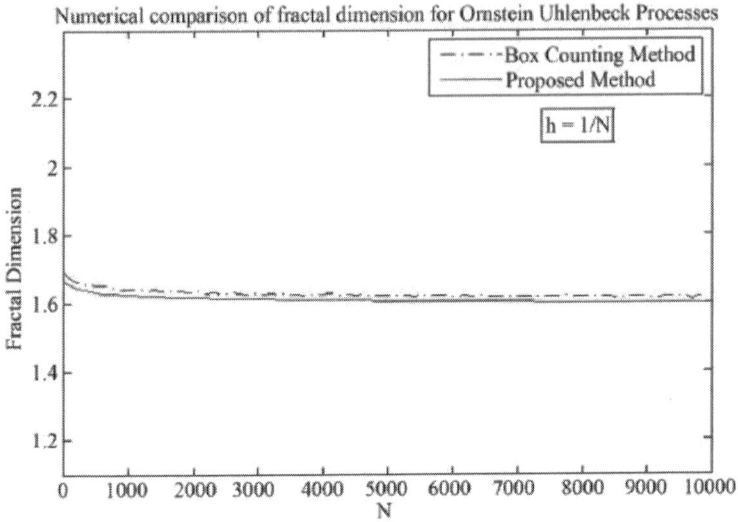

(a)

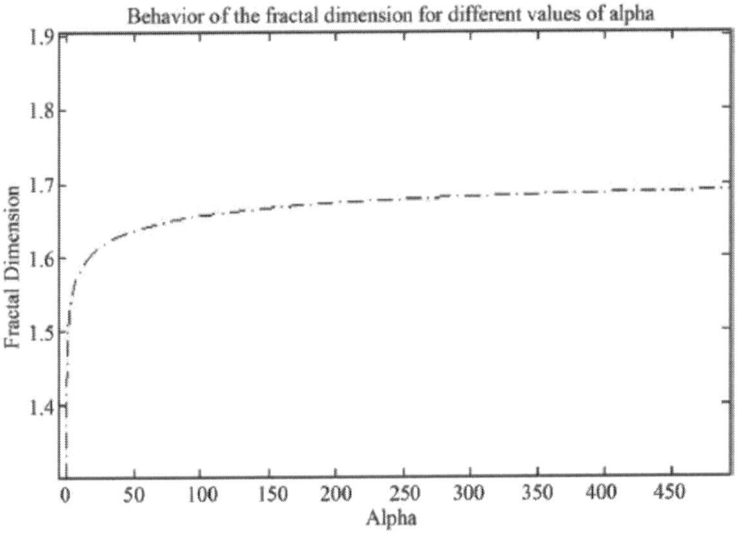

(b)

Figure 1. (a) Numerical comparison of fractal dimension for Ornstein Uhlenbeck processes with $T = 1$, $\alpha = 10$ and $\sigma = 20$ for different values of h; (b) Dynamic behavior of the fractal dimension for Ornstein Uhlenbeck processes with $T = 1$, $h = 1/10^4$ and $\sigma = 20$ for $0.001 \leq \alpha \leq 500$.

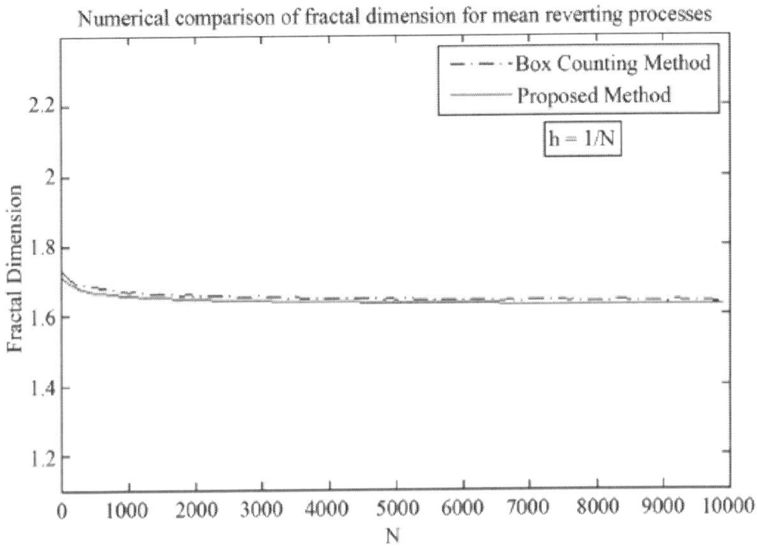

(a)

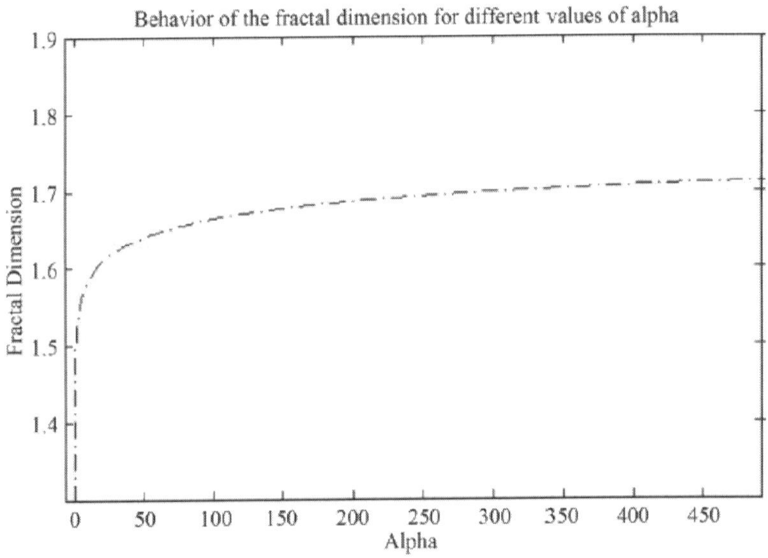

(a)

Figure 2. (a) Numerical comparison of fractal dimension for mean reverting processes with $T = 1$, $\mu = 10$, $\alpha = 20$and $\sigma = 250$ for different values of h; (b) Dynamic behavior of the fractal dimension for mean reverting processes with $T = 1$,, $\mu = 10$, $h = 1/10^4$ and $\sigma = 500$ for $0.001 \leq \alpha \leq 500$.

REFERENCES

1. Richardson, L. (1961) The Problem of Contiguity: An Appendix of Statistic of Deadly Quarrels. General Systems Year Book, 61, 139-187.

2. Mandelbrot, B. (1967) How Long Is the Coast of Britain Statistical Self-Similarity and Fractional Dimension. Science, 156, 636-638.

3. Duque, J.C., Betancourt, A. and Marin, F. (2014) An Algorithmic Approach for Simulating Realistic Irregular Lattices. In: Thill, J.C. and Dragicevic, S., Eds., GeoComputational Analysis and Modeling of Regional Systems, Springer Heidelberg, in press.

4. Mandelbrot, B. (1975) Stochastic Models for the Earth's Relief, the Shape and the Fractal Dimension of the Coastlines, and the Number-Area Rule for Islands. Proceedings of the National Academy of Sciences, 72, 3825-3828.

5. Adler, R.J. (1978) Some Erratic Patterns Generated by the Planar Wiener Process. Advances in Applied Probability, 10, 22-27.

6. Cullin, W.E. and Datko, M. (1987) The Fractal Geometry of the Soil-Covered Landscape. Earth Surface Processes and Landforms, 12, 369-385.

7. Cullin, W.E. (1988) Dimension and Entropy in the Soil-Covered Landscape. Earth Surface Processes and Landforms, 13, 619-648.

8. Beauvais, A. Montgomery, D. (1996) Influence of Valley Type on the Scaling Properties of River Planforms. Water Resources Research, 32, 1441-1448.

9. Montgomery, K. (1996) Sinuosity and Fractal Dimension of Meandering Rivers. Area, 28, 491-500.

10. Theiler, J. (1990) Estimating Fractal Dimension. JOSA A, 7, 1055-1073.

11. Taylor C. and Taylor, S.J. (1991) Estimating the Dimension of a Fractal. Journal of the Royal Statistical Society. Series B (Methodological), 53, 353-364.

12. Allen, M., Brown, G.J. and Miles, N.J. (1995) Measurement of Boundary Fractal Dimensions: Review of Current Techniques. Powder Technology, 84, 1-14.

13. Blachowski, A. and Ruebenbauer, K. (2009) Roughness Method to Estimate Fractal Dimension. Acta Physica Polonica A, 115, 636-640.

14. Girault, J., Kouame, D. and Ouahabi, A. (2010) Analytical Formulation of the Fractal Dimension of Filtered Stochastic Signals. Signal Processing, 90, 2690-2697.

15. Loeve, M. (1963) Probability Theory. D. Van N. Company, London.

16. Soong, T. (1973) Random Differential Equations in Science and Engineering. Academic Press, New York.

17. Jodar, L., Cortez, J. and Villafuerte, L. (2007) Mean Square Numerical Solution of Random Differential Equations: Facts and Possibilities. Computers and Mathematics with Applications, 53, 1098-1106.

18. Jodar, L. and Villafuerte, L. (2007) Numerical Solution of Random Differential Equations: A Mean Square Approach. Mathematical and Computer Modelling, 45, 757-765.

19. Marin, F. and Laniado, H. (2014) Convergence of Analytical Stochastic Processes in Mean Square. 1-9.

20. Agarwall, R. (1992) Difference Equations and Inequalities. Marcel Dekker, New York.

21. Barnsley, M. (1988) Fractals Everywhere. Academic Press, London.

22. Le Tavernier, E. (1998) La methode de Higuchi pour la dimension fractal. Signal Processing, 65, 115-128.

23. Uhlenbeck, G. and Ornstein, L. (1930) On the Theory of the Brownian Motion. Physical Review, 36, 823-841.

24. Marin, F. and Palacio, J. (2013) Gaussian Estimation of One-Factor Mean Reversion Processes. Journal of Probability and Statistics, 2013, 10 p.

NOTES

[1] See for details of second-order random variables (2-r.v.), second-order stochastic processes (2-s.p.), and random differential equations.

[2] Some convergence criteria random Euler scheme in the mean square sense can be viewed at -.

[3] The notation $f(\epsilon) \approx g(\epsilon)$ as follows. $\lim_{\epsilon \to 0} \{\ln(f(\epsilon))/\ln(g(\epsilon))\} = 1$, where $f(\epsilon)$ and $G(\epsilon)$ be real valued functions of the positive real variable ϵ.

CHAPTER 6

Stochastic Approximation Method for Fixed Point Problems

Ya. I. Alber[1], C. E. Chidume[2], Jinlu Li[3]

[1] Department of Mathematics, The Technion-Israel Institute of Technology, Haifa, Israel

[2] The Abdus Salam International Centre for Theoretical Physics, Trieste, Italy

[3] Department of Mathematical Sciences, Shawnee State University, Portsmouth, USA

ABSTRACT

We study iterative processes of stochastic approximation for finding fixed points of weakly contractive and nonexpansive operators in Hilbert spaces under the condition that operators are given with random errors. We prove mean square convergence and convergence almost sure (a.s.) of iterative approximations and establish both asymptotic and nonasymptotic estimates of the convergence rate in degenerate and non-degenerate cases. Previously the stochastic approximation algorithms were studied mainly for optimization problems.

KEYWORDS

Hilbert Spaces; Stochastic Approximation Algorithm; Weakly Contractive Operators; Nonexpansive Operators; Fixed Points; Convergence in Mean Square; Convergence Almost Sure (a.s.); Nonasymptotic Estimates of Convergence Rate

1. INTRODUCTION

In this paper the following problem is solved: To find a fixed point x^* of the operator $T: G \to H$, in other words, to find a solution $x^* \in G$ of the equation

$$x = Tx, \tag{1.1}$$

where T is a Lipschitz continuous mapping, H is a Hilbert space, $G \subseteq H$ is a closed convex subset. We suppose that x^* exists, i.e., the fixed point set N of T is nonempty. Note in different particular cases of the Equation (1.1), for example, when $T: G \to G$, the solution existence and solution uniqueness can be proved under some additional assumptions.

We separately consider two classes of mappings T: the class of weakly contractive maps and more general class of nonexpansive ones. Let us recall their definitions.

Definition 1.1. A mapping $T: G \to H$ is said to be weakly contractive of class $C_{\phi(t)}$ on a closed convex subset $G \subset H$ if there exists a continuous and increasing function $\phi(t)$ defined on IR^+ such that ϕ is positive on $IR^+/\{0\}, \phi(0) = 0, \lim_{t \to +\infty} \phi(t) = +\infty,$ and for all

$$x, y \in G, \|Tx - Ty\| \le \|x - y\| - \phi(\|x - y\|). \tag{1.2}$$

Remark 1.2. It follows from (1.2) that $\psi(t) \leq t$ and in real problems an argument t of the function $\psi(t)$ doesn't necessary approaches to ∞ obeying the condition $T: G \to H$ (see the example in Remark 3.4).

Definition 1.3. A mapping $T: G \to H$ is said to be nonexpansive on the closed convex subset $G \subset H$ if for all $x, y \in G$

$$\|Tx - Ty\| \leq \|x - y\|.$$

It is obvious that the class of weakly contractive maps is contained in the class of nonexpansive maps because the right-hand side of (1.2) is estimated as

$$0 \leq \|x - y\| - \phi(\|x - y\|) \leq \|x - y\|, \tag{1.3}$$

and it contains the class of strongly contractive maps because $\phi(t) = (1 - q)t$ with $0 < q < 1$ gives us

$$\|Tx - Ty\| \leq q\|x - y\|. \tag{1.4}$$

We study the following algorithm of stochastic approximation:

$$x_{n+1} = Pr_G\left(x_n - \alpha_n S_n x_n\right), n = 1, 2, \cdots, x_1 \in G, \tag{1.5}$$

where Pr_G is the metric projection operator from H onto G and deterministic step-parameters α_n satisfy the standard conditions:

$$\sum_{n=1}^{\infty} \alpha_n = \infty \text{ and } \sum_{n=1}^{\infty} \alpha_n^2 < \infty. \tag{1.6}$$

The factor $S_n x_n$ in (1.5) is an infinite-dimensional vector of random observations of the clearance operator $F = I - T$ at random points $x_n \in G$ given for all $n \geq 1$ on the same probability space (Ω, A, P). We set

$$S_n x_n = \zeta_n + \xi_n, \tag{1.7}$$

where $\zeta_n = x_n - Tx_n$ and ξ_n is a sequence of independent random vectors with the conditions

$$E[\xi_n] = 0 \text{ and } E\left[\|\xi_n\|^2\right] \le C_1, 0 < C_1 < \infty. \tag{1.8}$$

Here E is a symbol of the mathematical expectation. In order to calculate conditional mathematical expectations of different random variables we define the σ - subalgebra $A_n := \sigma(x_1, x_2, \cdots, x_n)$ on (Ω, A, P). And then $E[\xi_n / A_n]$ means A_n-measurable function with the following property: for any $B \in A_n$

$$\int_B \xi_n \mathrm{d}P(\omega) = \int_B E[\xi_n / A_n] \mathrm{d}P(\omega).$$

We also assume in the sequel that ζ_n is A_n-measurable for all $n \ge 1$.

Let us recall the mean square convergence and almost sure (a.s.) convergence.

We say that the sequence $\{\xi_n\}$ of random variables $\xi_n(\omega)$ converges in mean square to ξ if ξ exists and

$$\lim_{n\to\infty} E\left[\|\xi_n - \xi\|^2\right] = 0.$$

The sequence ξ_n converges to ξ almost surely or with probability 1 if

$$P\left(\lim_{n\to\infty} \xi_n(\omega) = \xi(\omega)\right) = 1.$$

Almost sure convergence and convergence in mean square imply convergence in the sense of probability: The sequence $\{\xi_n\}$ of random variables $\xi_n(\omega)$ converges in the sense of probability to $\xi(\omega)$ if for all $\varepsilon > 0$

$$\lim_{n\to\infty} P\left(\|\xi_n(\omega) - \xi(\omega)\| \ge \varepsilon\right) = 0.$$

So, we consider iterative processes of stochastic approximation in the form (1.5) for finding fixed points of weakly contractive (Definition 1.1) and nonexpansive (Definition 1.3) mappings in Hilbert spaces under the conditions (1.8). We prove mean square convergence and convergence almost sure of iterative approximations and establish both asymptotic and nonasymptotic estimates of the convergence rate. Perhaps, we present here the first results of this sort for fixed point problems. Formerly the stochastic approximation methods were studied mainly to find minimal and maximal points in optimization problems (see, for example, [1-6] and references within).

2. AUXILIARY RECURRENT INEQUALITIES

Lemma 2.1. [3,4] Let $\{\mu_k\},\{\rho_k\}$ and $\{\gamma_k\}$ be sequences of nonnegative real numbers satisfying the recurrent inequality.

$$\mu_{k+1} \le (1+\rho_k)\mu_k + \gamma_k,\ k=1,2,\cdots \tag{2.1}$$

Assume that $\sum_{k=1}^{\infty}\rho_k < \infty$ and $\sum_{k-1}^{\infty}\gamma_k < \infty.$ Then $\{\mu_k\}$ is bounded and converges to some limit.

Lemma 2.2. [3,4] Let $\{\mu_k\},\{\zeta_k\},\{\alpha_k\}$ and $\{\gamma_k\}$ be sequences of nonnegative real numbers satisfying the recurrent inequality.

$$\mu_{k+1} \le (1+\rho_k)\mu_k - \alpha_k\psi(\mu_k) + \gamma_k,\ k=1,2,\cdots, \tag{2.2}$$

where $\sum_{k=1}^{\infty}\alpha_k = \infty, \sum_{k=1}^{\infty}\rho_k < \infty$ and either $\sum_{k-1}^{\infty}\gamma_k < \infty$ or

$$\lim_{k\to\infty}\frac{\gamma_k}{\alpha_k} = 0. \tag{2.3}$$

Assume that $\psi(t)$ is continuous and increasing function defined on IR^1 such that ψ is positive on $IR^1/\{0\}$, $\psi(0)=0$. Then $\lim_{k\to\infty}\mu_k=0$. There exists an infinite subsequence $\{k_l\}, l=1,2,\cdots,$ such that

$$\mu_{k_l} \le C_0\psi^{-1}\left(\frac{1}{\sum_{k=1}^{k_l}\alpha_k}+\frac{\gamma_{k_l}}{\alpha_{k_l}}\right),$$

where $C_0 \ge \prod_{k=1}^{\infty}(1+\rho_k)$.

In the following two lemmas we want to present nonasymptotic estimates for the whole sequence $\mu_k, k\ge 1$. For this the stronger requirements are made of parameters α_k and function $\psi(t)$ in the recurrent inequality.

Suppose that $\alpha(t)$ such that $\alpha(k)=\alpha_k$, $F(t)$ and $\Phi(t)$ are antiderivatives from $\alpha(t)$ and $\frac{1}{\psi(t)}$ respectively, with arbitrary constants C (without loss of generality, one can put $C=0$), i.e.

$$F(t)=\int\alpha(t)dt, \Phi(t)=\int\frac{dt}{\psi(t)}.$$

Observe that $F(t)$ has the following properties:

i) $F'(t)=\alpha(t)$;

ii) $F(t)$ is strictly increasing on $[1,\infty)$ and $F(t)\to\infty$ as $t\to\infty$;

iii) The function $g(t)=\frac{\alpha(t)}{F(t)}=\frac{F'(t)}{F(t)}$ is decreasing;

iv) $G(t)=\ln F(t)\to\infty$ as $t\to\infty$.

Introduce the following denotations:

1) $\psi^{-1}(z)$ and $\Phi^{-1}(z)$ are the inverse functions to $\psi(t)$ and $\Phi(t)$, respectively;

2) $v(s)=\psi^{-1}\left[\dfrac{c_0\gamma(s-1)}{\alpha(s-1)}\right]+\gamma(s-1), s\ge 2, c_0>1$ is a fixed control parameter;

3) $u(s,C)=\Phi^{-1}\left[\Phi(C)-a(F(s)-F(2))\right]$, $C\ge 0, a=\dfrac{c_0-1}{c_0}<1;$

4) $w(s,v(r))=\Phi^{-1}\left[\Phi(v(r))-a(F(s)-F(2))\right]$, $2\le r<\infty;$

5) $Q(\bar{c})=\Phi^{-1}\left[\Phi(\bar{c})-a(F(2)-F(1))\right]$,

where

$\bar{c}>0$ is an arbitrary fixed number;

We present now the based condition (P): The graphs of the scalar functions $v(s)$ and $w(s,v(r))$ with any fixed $r\in[2,\infty)$ are intersected on the interval $[2,\infty)$ not more than at two points S_1 and S_2 (we do not consider contact points as intersection ones excepting $s=2$ if any).

For example, the graphs of the functions $v(s)$ and $w(s,v(r))$ calculated for $\alpha(s)=\dfrac{b}{s^{\kappa}}, b>0, 0<\kappa\le 1,$ and $\psi(t)=t^{\nu}, \nu\ge 1,$ satisfy the condition (P).

Lemma 2.3. [3,4] Assume that 1) the property (P) is carried out for the function $w(s,v(2))=u(s,v(2))$ and $v(s)$; 2) $u(s,v(2))\ge v(s)$ as $s\to\infty$; 3) the control parameter c_0 is chosen such that

$$u(s,v(2))\ge v(s) \text{ as } s\to 2. \tag{2.4}$$

Then for the sequence $\{\mu_k\}$ generated by the inequality

$$\mu_{k+1} \le \mu_k - \alpha_k \psi(\mu_k) + \gamma_k, k = 1, 2, \cdots, \tag{2.5}$$

it follows: $\lim_{k\to\infty} \mu_k = 0$ and for all $k \ge 1$

$$\mu_k \le u(k, C), C = \max\{Q(\mu_1), v(2)\}. \tag{2.6}$$

Lemma 2.4. [3,4] Assume that 1) the property (P) is carried out for all the function $w(s, v(2r))$ and $v(s)$; 2) $u(s, v(2)) \le v(s)$ as $s \to \infty$. Then for the sequence $\{\mu_k\}$ generated by the inequality (2.5) $\lim_{k\to\infty}\mu_k = 0$. In additiona) if $Q(\mu_1) \le v(2)$ and the control parameter c_0 is chosen such that $u(s, v(2)) \le v(s)$ as $s \to 2$, then for all $k \ge 2$

$$\mu_k \le v(k); \tag{2.7}$$

b) in all remaining cases

$$\mu_k \le u(k, C), C = \max\{Q(\mu_1), v(2)\},$$
$$1 \le k \le \bar{s}, \tag{2.8}$$

$$\mu_k \le v(k), k > \bar{s}, \tag{2.9}$$

where $\bar{s}$ is a unique root of the equation

$$u(s, C) = v(s) \tag{2.10}$$

on the interval $[2, \infty)$.

The following lemmas deal with another sort of recurrent inequalities:

Lemma 2.5. [7,8] Let $\{\mu_k\}, \{\alpha_k\}, \{\beta_k\}$ and $\{\gamma_k\}$ be sequences of non-negative real numbers satisfying the recurrence inequality.

$$\mu_{k+1} \leq \mu_k - \alpha_k \beta_k + \gamma_k, k = 1, 2, \cdots. \tag{2.11}$$

Assume that

$$\sum_{k=1}^{\infty} \alpha_k = \infty \text{ and } \sum_{k=1}^{\infty} \gamma_k < \infty.$$

Then:

There exists an infinite subsequence $\{\beta_{\ell_k}\} \subset \{\beta_k\}$ such that

$$\beta_{\ell_k} \leq \frac{1}{\sum_{j=1}^{\ell_k} \alpha_j}, \tag{2.12}$$

and, consequently, $\lim_{k\to\infty} \beta_{\ell_k} = 0;$

ii) if $\lim_{k\to\infty} \alpha_k = 0$ and there exists $\kappa > 0$ such that

$$|\beta_{k+1} - \beta_k| \leq \kappa \alpha_k \tag{2.13}$$

for all $k \geq 1$, then $\lim_{k\to\infty} \beta_k = 0.$

Lemma 2.6. [7,8] Let $\{\mu_k\}, \{\alpha_k\}, \{\beta_k\}$ and $\{\gamma_k\}$ be sequences of non-negative real numbers satisfying the recurrence inequality (2.11). Assume that $\sum_{k=1}^{\infty} \alpha_k = \infty$ and (2.3) is satisfied. Then there exists an infinite subsequence $\{\beta_{\ell_k}\} \subset \{\beta_k\}$ such that $\lim_{k\to\infty} \beta_{\ell_k} = 0.$

3. MEAN SQUARE CONVERGENCE OF STOCHASTIC APPROXIMATIONS

Theorem 3.1. Assume that $T:G\to H$ is a weakly contractive mapping of the class $C_{\phi(\tau)}$, $\psi_1(\tau)=\tau\phi(\tau)$ is a convex function with respect to $t=\tau^2$ and $E\left[\left\|x_1-x^*\right\|^2\right]<\infty$. Then the sequence $\{x_n\}$ generated by (1.5)-(1.7) converges in mean square to a unique fixed point x^* of T. There exists an infinite subsequence $\{n_l\}, l=1,2,\cdots,$ such that

$$E\left[\left\|x_{n_l}-x^*\right\|^2\right]\le C_0\psi^{-1}\left(\frac{1}{2\sum_{n=1}^{n_l}\alpha_n}+C_1\alpha_{n_l}\right), \tag{3.1}$$

where $\psi(\tau^2)=\psi_1(\tau)$ and some positive constant C_0 satisfies the inequality

$$\prod_{n=1}^{\infty}\left(1+8\alpha_n^2\right)\le C_0. \tag{3.2}$$

Remark 3.2. $1<C_0<\infty$ exists in view of the second condition in (1.6).

Proof. First of all, we note that the method (1.5)-(1.7) guarantees inclusion $x_n\in G$ for all $n\ge 1$. Since the metric projection operator Pr_G is nonexpansive in a Hilbert space and $x^*\in G$ exists, we can write

$$\begin{aligned}
&\left\|x_{n+1}-x^*\right\|^2\\
&=\left\|Pr_G\left(x_n-\alpha_n\left(x_n-Tx_n+\xi_n\right)\right)-Pr_G x^*\right\|^2\\
&\le\left\|x_n-x^*\right\|^2-2\alpha_n\left(\zeta_n+\xi_n,x_{n+1}-x^*\right)\\
&=\left\|x_n-x^*\right\|^2-2\alpha_n\left(\zeta_n,x_n-x^*\right)\\
&-2\alpha_n\left(\xi_n,x_n-x^*\right)+\left\|x_{n+1}-x_n\right\|^2.
\end{aligned} \tag{3.3}$$

Let us evaluate the first scalar product in (3.3). We have

$$\begin{aligned}
&\left(x_n - Tx_n, x_n - x^*\right)\\
&= \left(x_n - x^* - Tx_n + Tx^*, x_n - x^*\right)\\
&= \left\|x_n - x^*\right\|^2 - \left(Tx_n - Tx^*, x_n - x^*\right)\\
&\geq \left\|x_n - x^*\right\|^2 - \left\|Tx_n - Tx^*\right\|\left\|x_n - x^*\right\|\\
&\geq \left\|x_n - x^*\right\|^2 - \left(\left\|x_n - x^*\right\| - \phi\left(\left\|x_n - x^*\right\|\right)\right)\left\|x_n - x^*\right\|\\
&\geq \phi\left(\left\|x_n - x^*\right\|\right)\left\|x_n - x^*\right\| = \psi_1\left(\left\|x_n - x^*\right\|\right)
\end{aligned} \tag{3.4}$$

We remember that $\psi(\tau^2) = \psi_1(\tau)$. Then the inequalities (3.3) and (3.4) yield

$$\begin{aligned}
&\left\|x_{n+1} - x^*\right\|^2 \leq \left\|x_n - x^*\right\|^2 - 2\alpha_n \psi\left(\left\|x_n - x^*\right\|^2\right)\\
&-2\alpha_n\left(\xi_n, x_n - x^*\right) + \alpha_n^2\left\|\varsigma_n + \xi_n\right\|^2.
\end{aligned} \tag{3.5}$$

Applying the conditional expectation with respect to A_n to the both sides of (3.5) we obtain

$$\begin{aligned}
&E\left[\left\|x_{n+1} - x^*\right\|^2 / A_n\right]\\
&\leq \left\|x_n - x^*\right\|^2 - 2\alpha_n \psi\left(\left\|x_n - x^*\right\|^2\right)\\
&-2\alpha_n E\left[\left(\xi_n, x_n - x^*\right)/A_n\right] + \alpha_n^2 E\left[\left\|\varsigma_n + \xi_n\right\|^2 / A_n\right].
\end{aligned} \tag{3.6}$$

It is easy to see that

$$\begin{aligned}
&E\left[\left\|\varsigma_n + \xi_n^2\right\| / A_n\right]\\
&\leq 2E\left[\left\|\varsigma_n\right\|^2 / A_n\right] + 2E\left[\left\|\xi_n\right\|^2 / A_n\right]\\
&= 2\left\|\varsigma_n\right\|^2 + 2E\left[\left\|\xi_n\right\|^2 / A_n\right].
\end{aligned} \tag{3.7}$$

Since T is weakly contractive and therefore nonexpansive, one gets

$$\|x_n - Tx_n\|^2 \le 4\|x_n - x^*\|^2.$$

Taking into account (3.7), the inequality (3.9) is estimated as follows:

$$\begin{aligned} &E\left[\|x_{n+1} - x^*\|^2 / A_n\right] \\ &\le \left(1 + 8\alpha_n^2\right)\|x_n - x^*\|^2 - 2\alpha_n \psi\left(\|x_n - x^*\|^2\right) \\ &- 2\alpha_n E\left[\left(\xi_n, x_n - x^*\right)/A_n\right] + 2\alpha_n^2 E\left[\|\xi_n\|^2 / A_n\right]. \end{aligned} \tag{3.8}$$

Now the unconditional expectation implies

$$\begin{aligned} &E\left[\|x_{n+1} - x^*\|^2\right] \\ &\le \left(1 + 8\alpha_n^2\right)E\left[\|x_n - x^*\|^2\right] - 2\alpha_n E\left[\psi\left(\|x_n - x^*\|^2\right)\right] \\ &- 2\alpha_n E\left[\left(\xi_n, x_n - x^*\right)\right] + 2\alpha_n^2 E\left[\|\xi_n\|^2\right] \end{aligned} \tag{3.9}$$

Next we need the Jensen inequality for a convex function $\psi\left(\|x_n - x^*\|^2\right)$:

$$E\left[\psi\left(\|x_n - x^*\|^2\right)\right] \le \psi\left(E\left[\|x_n - x^*\|^2\right]\right)$$

(see [9,10]). This allows us to rewrite (3.9) in the form

$$\begin{aligned} &E\left[\|x_{n+1} - x^*\|^2\right] \\ &\le \left(1 + 8\alpha_n^2\right)E\left[\|x_n - x^*\|^2\right] \\ &- 2\alpha_n \psi\left(E\left[\|x_n - x^*\|^2\right]\right) + 2\alpha_n^2 E\left[\|\xi_n\|^2\right] \end{aligned} \tag{3.10}$$

because of

$$E\left[\left(\xi_n, x_n - x^*\right)\right] = 0.$$

Denoting $\lambda_n = E\left[\left\|x_n - x^*\right\|^2\right]$ we have

$$\lambda_{n+1} \leq \left(1+8\alpha_n^2\right)\lambda_n - 2\alpha_n\psi\left(\lambda_n\right) + 2C_1\alpha_n^2, \tag{3.11}$$

where in view of Definition 1.1, $\psi(\lambda)$ is a continuous and increasing function with $\psi(0)=0.$ Due to (6), from Lemma 2.2 it follows $\lim_{n\to\infty} E\left[\left\|x_n - x^*\right\|^2\right]=0$ and the estimate (3.1) holds too. The theorem is proved. □

Remark 3.3. If a fixed point of weakly contractive mapping $T:G\to H$ exists, then it is unique [11].

Remark 3.4. The following example was presented in [11]: Let $Tx = \sin x,\ G=[0,1]$ and $T:G\to G.$ It has been shown in [11] that

$$\left|\sin x - \sin y\right| \leq \left|x-y\right| - \frac{1}{8}\left|x-y\right|^3$$

for all $0\leq x\leq y\leq 1.$ Then

$$\phi(\tau)=\frac{1}{8}\tau^3, \psi_1(\tau)=\frac{1}{8}\tau^4 \text{ and } \psi(t)=\frac{1}{8}t^2.$$

Definition 3.5. Let a nonexpansive mapping

$T:G\to H$ have a unique fixed point x^*. T is said to be weakly sub-contractive on the closed convex subset $G\subset H$ if there exists continuous and increasing function $\psi(t)$ defined on IR^+ such that ψ is positive on

$IR^+/\{0\}$, $\psi(0)=0$, $\lim_{t\to+\infty}\phi(t)=+\infty,$ such that for all

$x\in G$.

$$\left\langle x-Tx, x-x^*\right\rangle \geq \psi\left(\left\|x-x^*\right\|^2\right). \tag{3.12}$$

Theorem 3.6. Assume that a mapping $T:G \to H$ is weakly sub-contractive and the function $\psi(t)$ in (3.12) is convex on IR^+ Then the results of Theorem 3.1 holds for the sequence $\{x_n\}$ generated by (1.5)-(1.7).

The second inequality in (1.6) can be omitted if we assume not less than linear growth of $\psi(\lambda)$"on infinity" and put $\alpha_n \to 0$ as $n \to \infty$

Theorem 3.7. Assume that a mapping $T:G \to H$ is weakly sub-contractive and the function $\psi(t)$ in (3.12) is convex on IR^+ Suppose that instead of (1.6) the conditions

$$\lim_{n\to\infty} \alpha_n = 0 \text{ and } \sum_{n=1}^{\infty} \alpha_n = \infty. \tag{3.13}$$

hold. In addition, let $\alpha_n \le \bar{\alpha} < 0.5$ and

$$\lim_{\lambda\to\infty} \frac{\psi(\lambda)}{\lambda} \ge 4\bar{\alpha}. \tag{3.14}$$

Then the sequence $\{x_n\}$ generated by (1.5), (1.7) and (3.13) converges in mean square to x^*. There exists an infinite subsequence $\{n_l\}, l = 1, 2, \cdots,$ such that

$$E\left[\left\|x_{n_l} - x^*\right\|^2\right] \le \psi^{-1}\left(\frac{1}{2\sum_{n=1}^{n_l} \alpha_n} + \frac{C_3 \alpha_{n_l}}{2}\right).$$

where

$$\begin{aligned} C_3 &= 2C_1 + 8\max\left\{\left\|x^* - x_1\right\|^2, C_2\right\}, \\ C_2 &= \lambda_* + 2C_1\bar{\alpha}^2 + 8\bar{\alpha}^2\lambda_*, \\ \lambda_* &= \psi^{-1}\left(C_1\bar{\alpha} + 4\bar{\alpha}\lambda_*\right). \end{aligned} \tag{3.15}$$

Proof. Consider the inequality (11) in the form

$$\lambda_{n+1} \le \lambda_n - 2\alpha_n \psi(\lambda_n) + 2C_1\alpha_n^2 + 8\alpha_n^2\lambda_n, \tag{3.16}$$

where $\lambda_n = E\left[\left\|x_n - x^*\right\|^2\right].$ Observe that it is derived by making use of (3.4) and the nonexpansivity property of $T.$ We shall show that λ_n are bounded for all $n \in I = \{1,2,\cdots\}.$ Indeed, since $\psi(\lambda)$ is a convex increasing continuous function, we conclude that $\psi_1(\lambda) = \frac{\psi(\lambda)}{\lambda}$ is nondecreasing and since (3.14) holdsthe inequality $\psi(\lambda) \le C_1\bar{\alpha} + 4\bar{\alpha}\lambda$ has a solution $\lambda \le \lambda_*,$ where λ_* is the unique root of the scalar equ0 ation $\psi(\lambda) = C_1\bar{\alpha} + 4\bar{\alpha}\lambda.$ Together with this, (3.4) and (3.14) are co-ordinated by the parameter $\bar{\alpha} < 0.5.$

Only one alternative can happen for each $n \in I$: either

$$H_1 := -2\alpha_n \psi(\lambda_n) + 2C_1\alpha_n^2 + 8\alpha_n^2\lambda_n \ge 0$$

or

$$H_2 := -2\alpha_n \psi(\lambda_n) + 2C_1\alpha_n^2 + 8\alpha_n^2\lambda_n < 0.$$

Denote $I_1 = \{n \in I \mid H_1 \text{ is true}\}$ and

$I_2 = \{n \in I \mid H_2 \text{ is true}\}$. It is clear that $I_1 \cup I_2 = I.$ From the hypothesis $H_1,$ it arises $\psi(\lambda_n) \le C_1\bar{\alpha} + 4\bar{\alpha}\lambda_n,$

and then $\lambda_n \le \lambda_*$ for all $n \in I_1.$ From the hypothesis $H_2,$ we have: $\lambda_{n+1} < \lambda_n$ for all $n \in I_2.$ Consider all the possible cases:

1) $I_2 = \varnothing.$ Then $\lambda_n \le \lambda_*$ for all $n \in I.$

2) $I_1 = \varnothing.$ Then $\lambda_n \le \lambda_1$ for all $n \in I.$

3) Let $I_1 = \{1, 2, \cdots, N_0\}$ and $I_2 = \{N_0 + 1, N_0 + 2, \cdots\}$. Then $\lambda_n \le \lambda_*$ for $n = 1, 2, \cdots, N_0$. By (3.16), $\lambda_{N_0+1} \le C_2$. It is obvious that $\lambda_n \le C_2$ for $N_0 + 2, N_0 + 3, \cdots$. Therefore, $\lambda_n \le C_2$ for all $n \in I$.

4) Let $I_2 = \{1, 2, \cdots, N_0\}$ and $I_1 = \{N_0 + 1, N_0 + 2, \cdots\}$. Then $\lambda_n \le \lambda_1$ for $n = 1, 2, \cdots, N_0$. and $\lambda_n \le \lambda_*$ for $N_0 + 1, N_0 + 2, \cdots$ Thus, $\lambda_n \le \lambda_*$ for all $n \in I$.

5) Let I_1 and I_2 be unbounded sets. Consider an arbitrary interval

$$I_s = [n_s + 1, n_{s+1} - 1] \subset I_2,$$

where $n_s, n_{s+1} \in I_1, s = 1, 3, 5, \cdots$. It is easy to be sure that $\lambda_{n_s} \le \lambda_*$ and $\lambda_n \le C_2$ for all $n \in I_s$.

6) The other situations of bounded and unbounded sets I_1 and I_2 are covered by the items 1)-5). Consequently, we have the final result: $\lambda_n \le \max\{\lambda_1, C_2\}$ for all $n \in I$.

Thus, we obtain the inequality

$$\lambda_{n+1} \le \lambda_n - 2\alpha_n \psi(\lambda_n) + C_3 \alpha_n^2, \quad (3.17)$$

where C_3 is defined by (3.15). Now Lemma 2.2 with the condition (2.3) implies the result. □

Remark 3.8. For a linear function $\psi(\tau) = c\tau$ which is convex and concave at the same time we suppose $2 \ge c > 4\bar{\alpha}$.

Remark 3.9. If G is bounded or more generally $\{x_n\}$ is bounded, then the inequality (3.17) (with some different constant C_3) immediately follows from (3.16).

4. ESTIMATES OF THE MEAN SQUARE CONVERGENCE RATE

Using Lemmas 2.3 and 2.4 we are able to give two general theorems on the nonasymptotic estimates of the mean square convergence rate for sequence $\{x_n\}$ generated by the stochastic approximation algorithm (1.5)-(1.7).

Again we introduce denotations 1)-5) from Section 2 induced now by the recurrent inequality (3.11):

1) $\psi^{-1}(z)$ and $\Phi^{-1}(z)$ are the inverse functions to $\psi(t)$ and $\Phi(t)$, respectively;

2) $v(s)=\psi^{-1}\left[C_1 c_0 \alpha(s-1)\right]+2C_1\alpha^2(s-1), c_0>1$ is a fixed control parameter;

3) $u(s,C_2)=\Phi^{-1}\left[\Phi(C_2)-a\left(F(s)-F(2)\right)\right], C_2>0,$

$a=\frac{c_0-1}{c_0}<1$

4) $w(s,v(r))=\Phi^{-1}\left[\Phi(v(r))-a\left(F(s)-F(2)\right)\right],$

$2\le s,r<\infty;$

5) $Q=\Phi^{-1}\left[\Phi\left(\|x_1-x^*\|^2\right)-a\left(F(2)-F(1)\right)\right].$

Introduce also the basic condition (P).

Theorem 4.1. Assume that all the conditions of Theorem 3.1 are fulfiled and

i) the condition (P) holds for the functions

$w(s,v(2))=u(s,v(2))$ and $v(s)$;

ii) $u(s,v(2)) \geq v(s)$ as $s \to \infty$;

iii) $c_0 > 1$ is chosen such that $u(s,v(2)) \geq v(s)$ as $s \to 2$.

Then the sequence $\{x_n\}$ generated by (1.5)-(1.7) converges in average to a unique fixed point x^* of T and for all $n \geq 1$

$$E\left[\left\|x_n - x^*\right\|^2\right] \leq C_0 u(n, C_2),$$
$$C_2 = \max\{Q, v(2)\}. \tag{4.1}$$

Theorem 4.2. Assume that all the conditions of Theorem 3.1 are fulfiled and

i) the condition (P) holds for the functions $v(s)$

and $w(s,v(z))$ with any fixed $z \in [2,\infty)$;

ii) $u(s,v(2)) \leq v(s)$ as $s \to \infty$;

iii) If $Q \leq v(2)$ and $c_0 > 1$ is chosen such that $u(s,v(2)) \leq v(s)$ as $s \to 2$, then the sequence $\{x_n\}$ generated by (1.5)-(1.7) converges in average to a unique fixed point x^* of T and for all $n \geq 1$

$$E\left[\left\|x_n - x^*\right\|^2\right] \leq C_0 v(n); \tag{4.2}$$

iv) In all the remaining cases, (4.1) holds for $1 \leq n \leq \bar{s}$ and (4.2) for $n \geq \bar{s}$, where $\bar{s}$ is a unique root of the equation $u(s, C_2) = v(s)$ on the interval $[2,\infty)$.

Let us provide the examples of functions $\psi(\tau)$ and $\Phi(\lambda)$ suitable for Theorems 4.1 and 4.2 (see [12,13]).

1) Below in Corollaries 4.3-4.6 we use the functions $\psi(\tau)=\tau^{\nu}$ with $\nu \geq 1$. For them

$$\Phi(\lambda)=\begin{cases}\ln\lambda, \text{ if } \nu=1,\\ \dfrac{\lambda^{1-\nu}}{1-\nu}, \text{ if } \nu\neq 1,\end{cases} \tag{4.3}$$

And

$$\Phi^{-1}(z)=\begin{cases}\exp z, \text{ if } \nu=1,\\ \left[(1-\nu)z\right]^{\frac{1}{1-\nu}}, \text{ if } \nu\neq 1,\end{cases} \tag{4.4}$$

2) If $\psi(\tau)=\exp\tau-1, \tau>0,$ then

$$\Phi(\tau)=\ln(1-\exp\tau) \text{ and } \Phi^{-1}(z)=-\ln(1-\exp z).$$

3) If $\psi(\tau)=\dfrac{\tau}{\tau+1}, \tau>0,$ then

$$\Phi(\tau)=\ln\tau+\tau \text{ and } \Phi^{-1}(z)\leq\ln(1+\exp z).$$

4) If $\psi(\tau)=\dfrac{\tau^2}{\tau+1}, \tau>0,$ then

$$\Phi(\tau)=\ln\tau-\frac{1}{\tau}.$$

In this example we are unable to define $\Phi^{-1}(z)$ in analitical form, therefore suggest to calculate it numerically by computer.

We next present very important corollaries from Theorems 4.1 and 4.2, where their assumptions automatically guarantee accomplishment of the condition (P) (see [4]). The functions $\psi(\tau)$ coincide with the point 1) above.

Corollary 4.3. Assume that $T:G\to H$ is a strongly contractive mapping, that is, (1.4) is satisfied with

$0<q<1$. Let in (1.5) $\alpha_n=\frac{b}{n}, b>0.$ Then

$$F(t)=b\ln t, \Phi(\tau)=\frac{\ln\tau}{q_1}, \Phi^{-1}(z)=e^{q_1 z}, q_1=1-q,$$

$$v(s)=\left(\frac{2c_0C_1}{b}+2C_1\right)\frac{1}{s-1}, u(s,C)=C\left(\frac{2}{s}\right)^{abq_1},$$

$$Q=\|x_1-x^*\|^2\left(\frac{1}{2}\right)^{abq_1}.$$

I. Suppose that $b>\frac{1}{q_1}$ and $c_0\geq\frac{bq_1}{bq_1-1}.$ Then

$\lim_{n\to\infty}E\left[\|x_n-x^*\|^2\right]=0$ and 1) If $Q\leq v(2)$ and $c_0\geq\frac{bq_1}{bq_1-2},$ we have for all $n>1$

$$E\left[\|x_n-x^*\|^2\right]\leq v(n); \tag{4.5}$$

In all the remain cases

$$E\left[\|x_n-x^*\|^2\right]\leq u(n,C),$$
$$C=\max\{Q,v(2)\}, 1<n\leq\bar{s}, \tag{4.6}$$

And

$$E\left[\|x_n-x^*\|^2\right]\leq v(s), n\geq\bar{s}, \tag{4.7}$$

where $\bar{s}$ is a unique root of the equation

$u(s,C)=v(s)$ on the interval $[2,\infty)$.

II. Suppose that $\frac{1}{2q_1} < b < \frac{1}{q_1}$ and $c_0 \geq \frac{bq_1}{1-bq_1} > 1$.

Then $\lim_{n\to\infty} E\left[\left\|x_n - x^*\right\|^2\right] = 0$ and the estimate (4.6) holds for all $n \geq 1$.

Corollary 4.4. Assume that $T : G \to H$ is a strongly contractive mapping, that is, in (1.4) is satisfied with $0 < q < 1$. Let in (1.5) $\alpha_n = \frac{b}{n^\kappa}, b > 0, 0 < \kappa < 1.$ Then

$$F(t) = \frac{b}{1-\kappa} t^{1-\kappa}, \Phi(\tau) = \frac{\ln\tau}{q_1},$$

$$\Phi^{-1}(z) = \left(\frac{1-\kappa}{b} z\right)^{\frac{1}{1-\kappa}}, q_1 = 1-q,$$

$$v(s) = \left(\frac{2c_0C_1}{b} + 2C_1\right)\frac{1}{s-1},$$

$$u(s,C) = C\exp\left\{-\frac{abq_1}{1-\kappa}\left(s^{1-\kappa} - 2^{1-\kappa}\right)\right\},$$

$$Q = \left\|x_1 - x^*\right\|^2 \exp\left\{-\frac{abq_1}{1-\kappa}\left(2^{1-\kappa} - 1\right)\right\}.$$

Suppose that $c_0 \geq \frac{bq}{bq-1}$. Then $\lim_{n\to\infty} E\left[\left\|x_n - x^*\right\|^2\right] = 0$

and

1) If $Q \leq v(2)$ and $c_0 \geq \frac{bq_1}{bq_1 - 2}$, we have for all $n > 1$

$$E\left[\left\|x_n - x^*\right\|^2\right] \leq v(n); \tag{4.8}$$

2) In all the remain cases the estimates (4.6) and (4.7) hold.

Corollary 4.5. Assume that $T : G \to H$ is a weakly contractive mapping of the class $C_{\phi(t)} = t^{2\nu-1}, \nu > 1,$ that is, in Theorem 3.1 $\psi(\tau) = \tau^\nu$. Let in (1.5)

$\alpha_n = \frac{b}{n}, b > 0.$ Then

$$F(t) = b\ln t, \Phi(\tau) = \frac{\tau^{1-\nu}}{1-\nu}, \Phi^{-1}(z) = \left[(1-\nu)z\right]^{\frac{1}{1-\nu}},$$

$$u(s,C) = C\left[1+(\nu-1)C^{\nu-1}ab\ln\frac{s}{2}\right]^{-\frac{1}{\nu-1}},$$

$$Q = \|x_1 - x^*\|^2\left[1+(\nu-1)\|x_1 - x^*\|^{2(\nu-1)} ab\ln 2\right]^{-\frac{1}{\nu-1}}.$$

If $c_0 > 1$ is chosen from the condition

$$\left(\frac{2c_0C_1}{b} + 2C_1\right)^{\nu-1} ab \le \nu^{-1},$$

then $\lim_{n\to\infty} E\left[\|x_n - x^*\|^2\right] = 0$ and for all $n > 1$

$$E\left[\|x_n - x^*\|^2\right] \le u(n,C),$$

$$C = \max\left\{Q, \left(\frac{2C_1c_0}{b}\right)^{\frac{1}{\nu}} + 2C_1\right\}. \tag{4.9}$$

Corollary 4.6. Assume that $T: G \to H$ is a weakly contractive mapping of the class $C_{\phi(t)} = t^{2\nu-1}, \nu > 1,$ that is, in Theorem 3.1 $\psi(\tau) = \tau^{\nu}$. Let in (1.5)

$\alpha_n = \frac{b}{n^{\kappa}}, b > 0, 0 < \kappa < 1.$ Then

$$F(t) = \frac{b}{1-\kappa}t^{1-\kappa}, \Phi(\tau) = \frac{\tau^{1-\nu}}{1-\nu}, \Phi^{-1}(z) = \left[(1-\nu)z\right]^{\frac{1}{1-\nu}},$$

$$v(s) = \left[\left(\frac{2c_0C_1}{b}\right)^{\frac{1}{\nu}} + 2C_1\right]\left(\frac{1}{s-1}\right)^{\frac{\kappa}{\nu}},$$

$$u(s,C) = C\left[1+\frac{\nu-1}{1-\kappa}C^{\nu-1}ab\left(s^{1-\kappa} - 2^{1-\kappa}\right)\right]^{-\frac{1}{\nu-1}},$$

$$Q = \|x_1 - x^*\|^2\left[1+\frac{\nu-1}{1-\kappa}\|x_1 - x^*\|^{2(\nu-1)} ab\left(2^{1-\kappa} - 1\right)\right]^{-\frac{1}{\nu-1}}.$$

I. Suppose that

$$\frac{\kappa}{\nu} < \frac{1-\kappa}{\nu-1}.$$

1) If $Q \le v(2)$ and $c_0 > 1$ is chosen from the condition

$$\left[\left(\frac{2c_0 C_1}{b}\right)^{\frac{1}{\nu}} + 2C_1\right]^{\nu-1} ab \ge \frac{2^{\kappa}\kappa}{\nu},$$

then $\lim_{n\to\infty} E\left[\left\|x_n - x^*\right\|^2\right] = 0$ and for all $n > 1$

$$E\left[\left\|x_n - x^*\right\|^2\right] \le v(n). \tag{4.10}$$

2) In all the remain cases the estimates (4.6) and (4.7) hold.

II. Suppose that

$$\frac{\kappa}{\nu} \ge \frac{1-\kappa}{\nu-1}.$$

If $c_0 > 1$ is chosen from the condition

$$\left[\left(\frac{2c_0 C_1}{b}\right)^{\frac{1}{\nu}} + 2C_1\right]^{\nu-1} ab \le \frac{\kappa}{\nu},$$

then $\lim_{n\to\infty} E\left[\left\|x_n - x^*\right\|^2\right] = 0$ and for all $n \ge 1$

$$E\left[\left\|x_n - x^*\right\|^2\right] \le u(n, C), C = \max\{Q, v(2)\}. \tag{4.11}$$

In addition to the examples presented in this section, we produce the functions $\psi(\tau)$ and $\Phi(\lambda)$ which have $\tau = 0$ as a tangency point of the infinite degree multiplicity and given logarithmic estimates of the convergence rate.

We define the function $\psi(\tau)$ by the following way:

$$\psi(\tau) = -\exp(-f(\tau))[f'(\tau)]^{-1}, \tau > 0,$$

where $f(\tau)$ is differentiable and decreasing function,

$\lim_{\tau \to 0} f(\tau) = \infty,$ and

$$\psi^{(l)}(0) = \lim_{\tau \to 0+} \psi^{(l)}(\tau), l = 0, 1, 2, \cdots,$$

where (l) denote the derivative degrees of the function $\psi(\tau), \psi^{(0)}(\tau) = \psi(\tau)$. It is easy to see that

$$\Phi(\tau) = -\exp f(\tau)$$

and

$$\Phi^{-1}(z) = f^{-1}(\ln(-z)).$$

In particular

i) $f(\tau) = \frac{1}{\tau}, \psi(\tau) = \tau^2 \exp\left(-\frac{1}{\tau}\right)$. We have

$\Phi(\tau) = -\exp\frac{1}{\tau}$ and $\Phi^{-1}(z) = \frac{1}{\ln(-z)}$. We have to verify that $\psi(\tau)$ is convex. In fact, it is true because

$$\frac{d^2\psi(\tau)}{d\tau} = \left(2 + \frac{2\tau + 1}{\tau^2}\right)\exp\left(-\frac{1}{\tau}\right) > 0, \forall \tau > 0.$$

Beside this, it is easy to see that $\psi(\tau) \le \tau$, at least, on the interval $[0,1]$. In the next examples we leave to readers to check these properties.

ii) $$f(\tau) = \frac{1}{\tau^s},\ s \ge 1,\ \psi(\tau) = \exp\left(-\frac{1}{\tau^s}\right)\tau^{s+1}.$$

We have $\Phi(\tau) = -\exp\frac{1}{\tau^s}$ and $\Phi^{-1}(z) = (\ln(-z))^{-\frac{1}{s}}$.

iii) $$f(\tau) = \exp\frac{1}{\tau^s}, \psi(\tau) = \exp\left(-\left[\frac{1}{\tau^s} + \exp\left(\frac{1}{\tau^s}\right)\right]\right)\tau^{s+1}$$

We have

$\Phi(\tau) = -\exp\left(\exp\frac{1}{\tau^s}\right)$ and $\Phi^{-1}(z) = (\ln\ln(-z))^{-\frac{1}{s}}$.

5. ALMOST SURE CONVERGENCE OF STOCHASTIC APPROXIMATIONS FOR NONEXPANSIVE MAPPINGS

Consider next the almost surely convergence of stochastic approximations. First of all, we need the stochastic analogy of Lemma 2.5:

Lemma 5.1. Let $\{\alpha_k\}$ be sequences of non-negative real numbers and $\{\beta_k\}$ be sequence of random A_k measurable variables, a.s. nonnegative for all $k \ge 1$. Assume that

$$\sum_{k=1}^{\infty} \alpha_k = \infty \text{ and } \sum_{k=1}^{\infty} \alpha_k \beta_k < \infty.$$

If $\lim_{k\to\infty} \alpha_k = 0$ and there exists $c > 0$ such that for all

$k \geq 1$

$$|\beta_{k+1} - E[\beta_k / A_n]| \leq c\alpha_k \text{ a.s.,} \tag{5.1}$$

then $\lim_{k\to\infty} \beta_k = 0$ a.s.

The proof can be provided by the scheme of nonstochastic case (see Proposition 2 in [8]) or as it was done in [5].

We need also the following lemma from [14] as applied to our case of Hilbert spaces (the concepts of modulus of convexity $\delta_B(\varepsilon)$ of Banach spaces B or Hilbert spaces H can be found in [15] and [16]).

Lemma 5.2. If $F = I - T$ with a nonexpansive mapping $T: G \to H$, then for all $x, y \in D(T)$,

$$(Fx - Fy, x - y) \geq R_1^2 \delta_H \left(\frac{\|Fx - Fy\|}{2R_1} \right),$$

where

$$R_1 = \sqrt{2^{-1}\left(\|x-y\|^2 + \|Tx - Ty\|^2\right)} \leq \|x - y\|.$$

If $\|x\| \leq R$ and $\|y\| \leq R$ with $x, y \in D(T)$, then $R_1 \leq 2R$ and

$$(Fx - Fy, x - y) \geq L^{-1} R^2 \delta_H \left(\frac{\|Fx - Fy\|}{4R} \right).$$

Theorem 5.3. Assume that a mapping $T: G \to H$ is nonexpansive and its fixed point set N is nonempty. If (1.8) holds and $E[\|\xi_n\| / A_n] \leq C_0$, then the sequence $\{x_n\}$ generated by (1.5)-(1.7) weakly almost surely converges to some $\tilde{x} \in N$.

Proof. Let $x^* \in N$. We next use Lemma 5.2 and the estimate (see [17], p. 49)

$$\delta_H(\varepsilon) \geq \frac{\varepsilon^2}{8}$$

to get

$$\left(x - Tx, x - x^*\right) \geq R_1^2 \delta_H \left(\frac{\|x - Tx\|}{2R_1} \right) \geq \frac{\|x - Tx\|^2}{32}.$$

In this case the inequality (3.3) implies

$$\begin{aligned} &\left\|x_{n+1} - x^*\right\|^2 \\ &\leq \left(1 + 8\alpha_n^2\right)\left\|x_n - x^*\right\|^2 - (32)^{-1}\left\|\alpha_n x_n - Tx_n\right\|^2 \\ &-2\alpha_n\left(\xi_n, x_n - x^*\right) + 2\alpha_n^2\left\|\xi_n\right\|^2. \end{aligned} \quad (5.2)$$

Similarly to (3.10), we have

$$\begin{aligned} &E\left[\left\|x_{n+1} - x^*\right\|^2 / A_n\right] \\ &\leq \left(1 + 8\alpha_n^2\right)\left\|x_n - x^*\right\|^2 - 2\alpha_n E\left[\left\|x_n - Tx_n\right\|^2 / A_n\right] \\ &- 2\alpha_n E\left[\left(\xi_n, x_n - x^*\right) / A_n\right] + 2\alpha_n^2 E\left[\left\|\xi_n\right\|^2 / A_n\right]. \end{aligned} \quad (5.3)$$

Denote $\lambda_n = E\left[\left\|x_n - x^*\right\|^2\right]$ and $\beta_n = E\left[\left\|x_n - Tx_n\right\|^2\right]$

and apply the unconditional expectation to both sides of (5.3). Then

$$\lambda_{n+1} \leq \left(1 + 8\alpha_n^2\right)\lambda_n - 2\alpha_n\beta_n + 2C_1\alpha_n^2. \quad (5.4)$$

It follows from this that

$$\lambda_{n+1} \leq \left(1 + 8\alpha_n^2\right)\lambda_n + 2C_1\alpha_n^2.$$

Since $\sum_1^\infty \alpha_n^2 < \infty$ and due to Lemma 2.1, we conclude that $\{\lambda_n\}$ is bounded. Consequently, $\{x_n\}$ is bounded a.s. that follows from the theory of convergent quasimartingales (see [5,18]).

We now need Lemma 5.1. It is not difficult to see that

$$\sum_1^\infty \alpha_n \beta_n = \sum_1^\infty \alpha_n E\left[\left\|x_n - Tx_n\right\|^2\right] < \infty. \tag{5.5}$$

The last gives us

$$\sum_1^\infty \alpha_n \left\|x_n - Tx_n\right\|^2 < \infty \text{ a.s.}$$

Next we evaluate the following difference:

$$\begin{aligned} &\left\|x_{n+1} - Tx_{n+1}\right\|^2 - \left\|x_n - Tx_n\right\|^2 \\ &\le \left(\left\|x_{n+1} - Tx_{n+1}\right\| + \left\|x_n - Tx_n\right\|\right) \\ &\times \left(\left\|x_{n+1} - Tx_{n+1}\right\| - \left\|x_n - Tx_n\right\|\right) \end{aligned}$$

It is easy to see that $\left\|x_n - Tx_n\right\|$ is bounded a.s. Indeed, since $\left\|x_n\right\| \le C_1$ a.s., there exists a constant $C_2 > 0$ such that

$$\left\|x_n - Tx_n\right\| = 2\left\|x_n - x^*\right\| \le C_2 \text{ a.s.}$$

Therefore

$$\left\|x_{n+1} - Tx_{n+1}\right\| + \left\|x_n - Tx_n\right\| \le 2C_2 \text{ a.s.}$$

It is obviously that

$$\|x_{n+1}-Tx_{n+1}\|-\|x_n-Tx_n\|$$
$$\leq\|(x_{n+1}-x_n)+(Tx_n-Tx_{n+1})\|\leq 2\|x_{n+1}-x_n\|$$
$$=2\|\Pr_G x_n-\alpha_n(x_n-Tx_n+\xi_n)-\Pr_G x_n\|$$
$$\leq 2\alpha_n\|x_n-Tx_n\|+2\alpha_n\|\xi_n\|.$$

Thus,

$$E\left[\|x_{n+1}-Tx_{n+1}\|^2/A_n\right]-\|x_n-Tx_n\|^2$$
$$\leq 4\alpha_n C_2^2+4\alpha_n C_2 E\left[\|\xi_n\|/A_n\right]$$
$$\leq 4C_2(C_2+C_0)\alpha_n \text{ a.s..}$$

By Lemma 5.1, $\|x_n-Tx_n\|\to 0$ a.s. as $n\to\infty$.

Since $\{x_n\}$ is bounded a.s., there is a subsequence $\{x_{n_k}\}$ weakly convergent to some point $\tilde{x}$. Since G is convex and closed, consequently, weakly closed, we assert that $\tilde{x}\in G$. It is known that a nonexpansive mapping T is weakly demiclosed, therefore, $\tilde{x}\in N$ a.s. Weak almost surely convergence of whole sequence $\{x_n\}$ is shown by the standard way [8].

Corollary 5.4. Assume that $T:G\to H$ is a weakly contractive mapping of the class $C_{\phi(t)}$. If $\sum_1^\infty \alpha_n=\infty$, $\sum_1^\infty \alpha_n^2<\infty$ and $E[\|\xi_n\|/A_n]\leq C_0$, then the sequence $\{x_n\}$ generated by (1.5)-(1.7) strongly almost surely converges to unique fixed point x^* of T.

Proof. We have from (3.4)

$$\|x_n-Tx_n\|\|x_n-x^*\|\geq\psi\left(\|x_n-x^*\|^2\right). \tag{5.6}$$

Since $\{x_n\}$ is bounded a.s. and $\|x_n-Tx_n\|\to 0$ a.s.

as $n \to \infty$, we conclude that $\psi\left(\left\|x_n - x^*\right\|^2\right) \to 0$ a.s.

The proof follows due to the properties of the function ψ. □

Remark 5.5. It is clear that all the results remain still valid for self-mappings $T: G \to G$. However, in this case, unlike any deterministic situation, the algorithm (1.5)-(1.7) must use the projection operator Pr_G because the vector $v_n = x_n - \alpha_n S_n x_n$ not always belongs to G. If $T_n x_n = Tx_n + \xi_n \in G$ for all $n \geq 1$ and $0 < \alpha_n < 1$, then (1.5) can be replaced by

$$x_{n+1} = (1-\alpha_n)x_n + \alpha_n T_n x_n, n = 1,2,\cdots, x_1 \in G.$$

REFERENCES

1. M. T. Wasan, "Stochastic Approximation," Cambridge University Press, Cambridge, 1969.

2. M. B. Nevelson and R. Z. Hasminsky, "Stochastic Approximation and Recursive Estimation," AMS Providence, Rhode Island, 1973.

3. Ya. Alber and S. Shilman, "Nonasymptotic Estimates of the Convergence Rate of Stochastic Iterative Algorithms," Automation and Remote Control, Vol. 42, 1981, pp. 32-41.

4. Ya. Alber and S. Shilman, "General Nonasymptotic Estimates of the Convergence Rate of Iterative Stochastic Algorithms," USSR Computational Mathematics and Mathematical Physics, Vol. 25, No. 2, 1985, pp. 13-20.

5. K. Barty, J.-S. Roy and C. Strugarek, "Hilbert-Valued Perturbed Subgradient Algorithms," Mathematics of Operation Research, Vol. 32, No. 3, 2007, pp. 551-562.

6. X. Chen and H. White, "Asymptotic Properties of Some Projection-Based Robbins-Monro Procedures in a Hilbert Space," Studies in Nonlinear Dynamics & Econometrics, Vol. 6, No. 1, 2002, pp. 1-53.

7. Ya. I. Alber and A. N. Iusem, "Extension of Subgradient Techniques for Nonsmooth Optimization in Banach Spaces," Set-Valued Analysis, Vol. 9, No. 4, 2001, pp. 315-335.

8. Ya. Alber, A. Iusem and M. Solodov, "On the Projection Subgradient Method for Nonsmooth Convex Optimization in a Hilbert Space," Mathematical Programming, Vol. 81, No. 1, 1998, pp. 23-35.

9. A. P. Dempster, N. M. Laird and D. B. Rubin, "Maximal Likelihood from Incomplete Data via the EM Algorithm," Journal of the Royal Statistical Society, Vol. 39, No. 1, 1977, pp. 185-197.

10. W. Rudin, "Real and Complex Analysis," McGraw-Hill, New York, 1978.

11. Ya. I. Alber and S. Guerre-Delabriere, "Principle of Weakly Contractive Maps in Hilbert Spaces," Operator Theory, Advances and Applications, Vol. 98, 1997, pp. 7- 22.

12. Ya. I. Alber, "Recurrence Relations and Variational Inequalities," Soviet Mathematics Doklady, Vol. 27, 1983, pp. 511-517.

13. Ya. Alber, S. Guerre-Delabriere and L. Zelenko, "The Principle of Weakly Contractive Maps in Metric Spaces," Communications on Applied Nonlinear Analysis, Vol. 5, No. 1, 1998, pp. 45-68.

14. Ya. I. Alber, "New Results in Fixed Point Theory," Preprint, Haifa Technion, 2000.

15. J. Diestel, "The Geometry of Banach Spaces," Lecture Notes in Mathematics, No. 485, Springer, Berlin, 1975.

16. T. Figiel, "On the Moduli of Convexity and Smoothness," Studia Mathematica, Vol. 56, No. 2, 1976, pp. 121-155.

17. Ya. Alber and I. Ryazantseva, "Nonlinear Ill-Posed Problems of Monotone Type," Springer, Berlin, 2006.

18. M. Métivier, "Semimartingales," De Gruyter, Berlin, 1982.

CHAPTER 7

CreditGrades Framework within Stochastic Covariance Models

Marcos Escobar[1], Hamidreza Arian[2], Luis Seco[3]

[1] Department of Mathematics, Ryerson University, Toronto, Canada

[2] Royal Bank of Canada, Toronto, Canada

[3] Department of Mathematics, University of Toronto, Toronto, Canada

ABSTRACT

In this paper we study a multivariate extension of a structural credit risk model, the CreditGrades model, under the assumption of stochastic volatility and correlation between the assets of the companies. The covariance of the assets follows two popular models which are non-overlapping extensions of the CIR model to dimensions greater than one, the Wishart process and the Principal component process. Under CreditGrades, we find quasi closed-form solutions for equity options, marginal probabilities of defaults, and some other major financial derivatives.

KEYWORDS

Barrier Option; Stochastic Covariance; CreditGrades Mode; Wishart Process; Principal Component Process

1. INTRODUCTION

We present a structural credit risk model which considers stochastic correlation between the assets of the companies. We allow the covariance of the assets to follow two popular stochastic covariance models. First we assume it follows a Wishart process [1] then we assume a Principal component Process (see [2]), both are non overlapping extensions of the CIR model to dimensions greater than one.

Modeling stochastic correlation has difficulties from the analytical as well as the estimation point of view. One of the attempts to x this gap began with a paper on Wishart processes by [1], which followed by a series of papers by Gourieroux, see [3]. Several authors have recently brought the finance community's attention to the Wishart process and showed that the Wishart process is a good candidate for modeling the covariance of assets. A Wishart process is an affine symmetric positive definite process. The stream of papers [3-7] brought the finance community's attention to this process as a natural extension of Heston's stochastic volatility model, which has been a very successful univariate model for option pricing and reconstruction of volatility smiles and skews. The popularity of the Heston model as well as the empirical evidence of stochastic correlation and volatility has contributed to the recent popularity of the Wishart process. Risk is usually measured by the covariance matrix. Therefore Wishart process can be seen as a tool to model dynamic behavior of multivariate risk. Via the Laplace transform and the distribution of the Wishart process [3], prices derivatives with a generalized Wishart stochastic covariance matrix. This approach can be used to model risk in the structural credit risk framework. DaFonseca extends his approach in [7] to model the multivariate risk by a Wishart process. In this paper several risky stocks are considered and the pricing problem for one dimensional vanilla options and multidimensional geometric basket options on the stocks is presented. We adapt existing results about the Wishart process to the structural CreditGrades framework. We give quasi closed-form solutions for equity options, marginal probabilities of defaults, and some other major financial derivatives. For calculation of our pricing formulas we make a bridge between two recent trends in pricing theory; from one side, pricing of barrier options by [8] and [9] and from other side the development of Wishart process by [1].

In the second part of the paper we develop a new model for credit risk based on a model with stochastic eigenvalues called principal component stochastic covariance. To induce the stochasticity into the structure of volatilities and correlation, we assume that the eigenvectors of the covariance matrix are constant but the eigenvalues are driven by independent Cox-Ingersoll-Ross processes. To price equity options on this framework we first transform the calculations from the pricing domain to the frequency domain. Then we derive a closed formula for the Fourier transform of the Green's function of the pricing PDE. Finally we use the method of images to find the price of the equity options. Same method is used to find closed formulas for marginal probabilities of defaults and CDS prices. Inspired by the standard stochastic volatility models starting from Heston's paper [10] and the work on [2] the applications of the principal component model in credit risk is studied. The main idea is to identify the covariance matrix by its eigenvectors and eigenvalues. In the above papers, authors have assumed that the eigenvector of the covariance matrix are constant but the eigenvalues follow a CIR process. This implies a stochastic structure for the correlation between the assets. [2] prices the collateralized debt obligations under the Merton's model using a tree approach and the principal component model.

Merton's model [11] is the first structural credit risk model proposed which considers the company's equity as an option on the firm's asset. There has been numerous extensions for the Merton's model in the literature including incorporating early defaults, stochastic interest rates, stochastic default barriers and jumps in the asset's price process. A simpler approach was jointly developed by CreditMetrics, JP Morgan, Goldman Sachs and Deutsche Bank, called the CreditGrades model, this can be seen as a particular case of Merton with zero time to maturity.

The simplicity allows for closed for expression on some derivatives as shown in our paper which can not be found closed form under Merton or Black Cox structural frameworks. We extend the CreditGrades model using stochastic covariance Wishart process focusing on the role of stochastic correlation. The performance of a company is usually monitored by observing its equity's volatility or the CDS spread. CreditGrades model can be considered as a down-and-out barrier credit risk model. This means that default is triggered if the value of the asset reaches a certain level identified by the recovery part of the debt. [12] has extended the CreditGrades model to price equity options by introducing the equity as a shifted log-normal process. [9] has extended [12] idea by embedding the Heston's volatility into the model and pricing equity derivative. Both [9] and [12] models are univariate credit risk models. We extend the CreditGrades model by use of the Wishart and PC processes to dimensions greater than one implementing stochastic correlation into the dynamics of the assets. We give quasi closed formulas for equity derivatives based on these models.

This paper is organized as follows: in section 2 we use Wishart process as a candidate to model the covariance matrix of the assets' prices within a CreditGrades model. The pricing problem for some derivatives on the equities is derived in section 2.2. Section 3 presents and uses the Principal component process for the covariance matrix of the assets' prices within the same structural framework. The pricing problem is derived in Section 3.2. Section 4 concludes. The proofs are given in the appendix.

2. THE CREDITGRADES WISHART PROCESS

In this section, we introduce our Wishart CreditGrades model. CreditGrades model is a version of the Merton model jointly developed by CreditMetrics, JP Morgan, Goldman Sachs and Deutsche Bank. The original version of the CreditGrades model assumes that volatility is deterministic. We extend CreditGrades model, by means of stochastic covariance Wishart process. Our model allows correlation and volatility be stochastic. By considering the stochastic covariance Wishart process, we have more flexibility and degree of freedom in the marginal, while analytic tractability is preserved when extending CIR process to Wishart process. We first present Wishart process of integer degree of freedom and derive their matrix stochastic differential equation which later on will give a natural representation of Wishart process with fractional degree of freedom. A Wishart process with integer degree of freedom K is a sum of K independent n-dimensional Ornstein-Uhlenbeck process. We first remind the formal definition below:

Definition: Consider $\{U^{(k)}\}_{k=1}^{K}$ as an independent set of Ornstein-Uhlenbeck processes

$$dU_t^{(k)} = AU_t^{(k)}dt + QdW_t^{(k)},$$

where A and Q are (n,n) matrices with Q invertible. Then a Wishart process of degree K is defined as

$$\Sigma_t = \sum_{k=1}^{K} U_t^{(k)} U_t^{(k)'},$$

where $U_t^{(k)'}$ is the transpose of the vector $U_t^{(k)}$.

Ito's lemma can be used to find a diffusion SDE for the process Σ_t

$$\begin{aligned} d\Sigma_t &= \sum_{k=1}^{K}\left(dU_t^{(k)}\cdot U_t^{(k)'} + U_t^{(k)}\cdot dU_t^{(k)'} + dU_t^{(k)}\cdot dU_t^{(k)'}\right) \\ &= \left(KQ'Q + A\Sigma_t + \Sigma_t A'\right)dt \\ &\quad + \sum_{k=1}^{K}\left(QdW_t^{(k)}U_t^{(k)'} + U_t^{(k)}dW_t^{(k)'}Q'\right). \end{aligned}$$

As it can be seen, the drift term of the SDE above contains Σ_t, but the diffusion part contains the terms $U_t^{(k)}$ and $U_t^{(k)'}$ separately. [1,3] show that Σ_t also satisfies the following matrix SDE

$$d\Sigma_t = \left(KQ'Q + A\Sigma_t + \Sigma_t A'\right)dt + QdW_t\Sigma_t^{\frac{1}{2}} + \Sigma_t^{\frac{1}{2}}dW_tQ',$$

where W_t is an $n \times n$ standard Brownian motion matrix.

2.1. The Dynamics of the Assets

The assets are defined on a probability space $(\Omega, \mathbf{F}, Q)$ where $\{\mathbf{F}_t\}_{t\geq 0}$ is the information up to time t and Q is the risk-neutral measure equivalent to the real-world measure P. Let's assume that the ith firm's asset price per share is given by $A_i(t)$. Here we review the results regarding the dynamics of the assets with stochastic covariance Wishart process. As before, assume the assets' prices follow the multivariate real-world model

$$\mathrm{d}\ln A_t = \left(\mu_i + \left(Tr\left(D_i\Sigma_t\right)\right)\right)\mathrm{d}t + \Sigma_t^{\frac{1}{2}}\mathrm{d}W_t^A,$$

$$\mathrm{d}\Sigma_t = \Omega\Omega' + M\Sigma_t + \Sigma_t M_t'^{\sigma}\Sigma_t^{\frac{1}{2}} + \Sigma_t^{\frac{1}{2}}\mathrm{d}W^{\sigma}Q'.$$

Here the vector $\mu = (\mu_1, \cdots, \mu_n)$ is constant and D_i is a symmetric positive definite matrix, the log-price process has the drift term $E_t(\mathrm{d}\ln A_t) = \left(\mu_i + \left(Tr\left(D_i\Sigma_t\right)\right)\right)\mathrm{d}t$ and the quadratic variation $V_t(\mathrm{d}\ln A_t) = \Sigma_t \mathrm{d}t$. Moreover, we assume that the Brownian motions driving the assets and the Brownian motions driving the Wishart process are uncorrelated. $Tr(D_i\Sigma_t) > 0$ accounts for risk premium (under risk neutral measure). For the transition distribution of A_{t+h} given A_t and (Σ_t) we have

$$\ln A_{t+h} \mid A_t, (\Sigma_t)$$
$$\sim \mathbf{N}\left(\ln A_t + \int_t^{t+h} \mu + Tr\left(D_i\Sigma_u\right)\mathrm{d}u, \int_t^{t+h} \Sigma_u \,\mathrm{d}u\right),$$

and the unconditional probability function can be found by Integration over the distribution function of $\int_t^{t+h} \Sigma_u \,\mathrm{d}u$.

Up to now we have described the dynamics of the assets' prices with stochastic covariance structure coming from the Wishart process. The asset's price for a firm is not directly observed from the market. This leads us to structural credit risk models which introduce the equity as a form of a derivative on the asset of the company. Then the role of the model is in connecting the equity market to the default event. For example [3] proposes a dynamics for the assets and liabilities in the Merton's model where the equity is defined as a call option on the asset with liability as the strike price. This is a direct extension of the Merton model with multivariate stochastic volatility. In that case the price of a bond has quasi closed-form formulas based on the closed formulas for the conditional Laplace transform of the joint price-covariance process. We prefer not to use this model for several reasons. First, we found the CreditGrades model more popular in financial markets because of its ability to link the structural framework to equity derivatives. On the other hand, the model proposed by [3] has one common Wishart process and one distinct Wishart processes for each asset. Therefore, the number of parameters for the model is high and the calibration is extremely illposed because of the high degree of freedom imposed by the number of parameters inside the model. We found that in the case of two companies, assuming only one Wishart process driving the covariance matrix gives a fairly flexible model to capture market's behavior, while at the same time provides a fewer number of parameters.

Next, we introduce the equity process based on the CreditGrades' perspective rather than Merton's perspective taking advantage of the flexibility of the CreditGrades model. This will enrich the credit risk modeling with possibility of early default and also a straightforward link between credit risk and the equity market. We will derive a formula for the infinitesimal generator of the joint equity-covariance process below. This operator will play an important role in the partial differential equation of the equity option's price.

Now that we have identified the dynamics of the assets, we explain the mechanism of the CreditGrades model. As before, we assume that the ith firm's value $A_i(t)$ is driven by the dynamics

$$\begin{cases} \mathrm{d}A_i(t) = \mathrm{diag}\left(A_i(t)\right)\left[\left(r(t) - d_i(t)\right) I\mathrm{d}t + \sqrt{\Sigma_t}\,\mathrm{d}W_t\right], \\ \mathrm{d}\Sigma_t = \left(\Omega\Omega' + M\Sigma_t + \Sigma_t M'\right)\mathrm{d}t + \sqrt{\Sigma_t}\,\mathrm{d}Z_t Q + Q_t''\sqrt{\Sigma_t}, \end{cases} \tag{1}$$

where $\Omega\Omega' = \beta Q'Q$ for some $\beta > n-1$ and M is a negative definite matrix. We assume that assets are driven by the Brownian motion W_t, the covariance matrix of the assets which follows a Wishart process is driven by the Brownian motion Z_t and two Brownian motions W_t and Z_t are uncorrelated. We assume that $S_i(t)$ is the ith firm's equity price per share, $B_i(t)$ is the ith firm's debt per share and R_i is the ith firm's recovery rate. $B_i(t)$ has a deterministic growth rate $r(t)-d_i(t)$, where $r(t)$ is the risk free interest rate and $d_i(t)$ is the dividend yield for the ith firm. The recovery part of the debt, $D_i(t) = R_i B_i(t)$, is the default barrier for the asset. Therefore, the default time based on CreditGrades model is given by

$$\eta_i = \inf\{0 \le t \,|\, A_i(t) \le D_i(t)\}.$$

In the framework of CreditGrades model, the equity's value is given by

$$S_i(t) = \begin{cases} A_i(t) - D_i(t), & t < \eta_i, \\ 0, & \eta_i \le t \end{cases}$$

In terms of the equity, the default time can be written as $\eta_i = \inf\{0 < t \,|\, S_i(t) \le 0\}$. Zero is an absorbing state for the equity process which makes the pricing of the equity option similar to pricing of down-and-out options studied by [11]. By using $S_i(t) = A_i(t) - D_i(t)$, the dynamics of $D_i(t)$ and equation (1), the equity follows a shifted log-normal SDE. We will use the notation $\operatorname{diag}(x)$, $vec(x)$ and I denoting diagonal matrix and vector with elements x and the matrix of ones respectively1.

$$\begin{cases} dS_i(t) = \operatorname{diag}\left(S_i(t) \cdot \left(r(t) - d_i(t)\right)\right) I dt \\ \qquad\qquad + \operatorname{diag}\left(S_i(t) + D_i(t)\right) \sqrt{\Sigma_t}\, dW_t, \\ d\Sigma_t = \left(\Omega\Omega' + M\Sigma_t + \Sigma_t M'\right) dt + \sqrt{\Sigma_t}\, dZ_t Q + Q_t' \sqrt{\Sigma_t} \end{cases} \tag{2}$$

Note that the solution of the dynamics above can reach negative values but not before the stopping time η_i. We force sufficient conditions on the Wishart process to make Σ_t mean reverting. For our purposes, we assume M is negative definite and $\Omega\Omega' = \beta Q'Q$ for some $\beta > n-1$. Moreover, without loss of generality, we assume $\Omega = \sqrt{\beta Q'}$. We first derive the infinitesimal generator of the joint process (S,Σ). This operator will appear in the pricing PDE for equity options and the probabilities of default in the next section Proposition 1: The infinitesimal generator of the joint process (S,Σ) is given by

$$\begin{aligned}\mathbf{A}_{(S,\Sigma)} &= \left[(r(t)-d(t))S\right]\nabla_S \\ &+\frac{1}{2}\left[(S+D)\nabla_S\right]\Sigma\left[(S+D)\nabla_S\right]' \\ &+Tr\left[(\Omega\Omega'+M\Sigma_t+\Sigma_t M')D+2\Sigma D Q'QD\right],\end{aligned}$$

where $D=\left(\frac{\partial}{\partial\Sigma_{ij}}\right)_{ij}$, Tr is the trace of a matrix, and we've used the notation $\left[(r-d)S\right]=Vec\left((r-d_i)S_i\right)$ and $\left[(S+D)\nabla_S\right]=Vec\left((S_i+D_i)\frac{\partial}{\partial S_i}\right)$.

2.2. Derivative Pricing; Analytical Results

In this section, we tackle the pricing problem of our credit risk model. We will use the fourier transform and method of images to solve the pricing problem for European calls and puts on the equity.

2.2.1. Equity Call Options

The price of a European Call option on the equity is calculated by discounting the risk-neutral expectation of the payoff at maturity. Since $\left(S_{i,T}1_{\{\eta_i>T\}}-K\right)^+=\left(S_{i,T}-K\right)^+1_{\{\eta>T\}}$ the price of the call option could be rewritten as

$$\begin{aligned}&V_{\text{call}}\left(t,\Sigma,S_i,K\right)\\ &=E^Q_{(t,\Sigma,S_i)}\left(\exp\left(-\int_t^T r(s)\,\mathrm{d}s\right)\left(S_{i,T}1_{\eta_i>T}-K\right)^+\right)\\ &=E^Q_{(t,\Sigma,S_i)}\left(\exp\left(-\int_t^T r(s)\,\mathrm{d}s\right)\left(S_{i,T}-K\right)^+1_{\eta_i>T}\right).\end{aligned}$$

The price of a single name derivative on one of the equities satisfies the partial differential equation $V_t + \mathbf{A}_{(\Sigma,S_i)} - rV = 0$. Specially, the price of an equity call option is given by the PDE

$$\begin{cases} V_t + \frac{1}{2}\Sigma_{ii}\left(S_i + D_i(t)\right)^2 V_{S_iS_i} + \left(r(t) - d_i(t)\right)S_i V_{S_i} \\ \quad + \mathbf{A}_\Sigma V - rV = 0, \\ V(t,0) = 0, V(T,S) = (S-K)^+, \end{cases} \tag{3}$$

where $\mathbf{A}_{(\Sigma,S)}$ is the infinitesimal generator of the joint process (S,Σ) given by the proposition 1. We first change the variables by $x_i = \ln\left(\frac{S_i + D_i(t)}{D_i(t)}\right)$, $a_i = \ln\left(\frac{D_i(T)+K}{D_i(T)}\right)$ and

$G(t,x_i) = \exp\left(\int_t^T r(s)\mathrm{d}s\right)\frac{V(t,S_i)}{D_i(T)}$ to transform the PDE to

$$\begin{cases} G_t + \frac{1}{2}\Sigma_{ii}\left(G_{x_ix_i} - G_{x_i}\right) + \mathbf{A}_\Sigma G = 0, \\ G(t,0) = 0, V(T,x_i) = \left(\mathrm{e}^{x_i} - \mathrm{e}^{a_i}\right)^+. \end{cases} \tag{4}$$

To use the method of images, we need to eliminate the drift term first, hence we change the variables by $y_i = x_i - a_i$, $\tau = T - t$ and $U(\tau, y_i) = \mathrm{e}^{-a}\mathrm{e}^{-\frac{y_i}{2}}G(t,x_i)$.

Then, PDE (3) transforms to

$$\begin{cases} -U_\tau + \frac{1}{2}\Sigma_{ii}U_{y_iy_i} + \mathbf{A}_\Sigma U - \frac{1}{8}\Sigma_{ii}U = 0, \\ U(\tau,0) = 0, U(0,y_i) = \left(\mathrm{e}^{\frac{y_i}{2}} - \mathrm{e}^{-\frac{y_i}{2}}\right)^+. \end{cases} \tag{5}$$

The PDE (5) is our reference PDE to solve the pricing problem for equity options on $S_i(t)$. We have the following proposition for the Fourier transform of the Green's function of PDE Proposition 2: The Fourier transform of the Green's function of PDE is given by

$$q_j(\tau,\Sigma,Y)=\int_{-\infty}^{+\infty}e^{ikY_j+A(\tau,k)+Tr(B(\tau,k)\Sigma)}\,dk, \tag{6}$$

where

$$B(\tau,k)=(\Lambda_{22}(\tau,k))^{-1}(\Lambda_{21}(\tau,k))$$
$$A(\tau,k)=Tr\left(\Omega\Omega'\int_0^\tau B(u,k)du\right)$$

With

$$\Lambda=\begin{pmatrix}\Lambda_{11} & \Lambda_{12}\\ \Lambda_{21} & \Lambda_{22}\end{pmatrix}=\exp\tau\begin{pmatrix}M & -2Q'Q\\ -\frac{1}{2}\left(k^2+\frac{1}{4}\right)I & -M'\end{pmatrix}.$$

Now that we have found the Fourier transform of the Green's function of the pricing PDE, we solve the pricing problem for an equity call option by the method of images.

Proposition 3: The price of a call option on $S_j(t)$ with maturity date T and strike price K is given by

$$V(t,S_j)=(D(T)+K)\exp\left(-\int_t^T r(s)ds\right)Z(\tau,y),$$

$$y=\ln(S+D(t))-\ln(D(T)+K)+\int_t^T(r(s)-d(s))ds$$
$$b=\ln(D(t))-\ln(D(T)+K)+\int_t^T(r(s)-d(s))ds$$

and the function Z is defined by

$$Z(\tau,y)=e^y-e^b$$
$$-\frac{e^{\frac{1}{2}y}}{\pi}\int_0^{+\infty}\frac{e^{A(\tau,k)+Tr(B(\tau,k)\Sigma)}\left(\cos(yk)-\cos((y-2b)k)\right)}{k^2+\frac{1}{4}}ds, \tag{7}$$

with $A(\tau,k)$ and $B(\tau,k)$ given as in proposition 2.

For large values of k, the integrand in (7) is exponentially decreasing which makes it easy to evaluate the integral numerically.

Remark 1: As we have mentioned before, our result covers [9] as a special case. If in the dynamics of the asset (1), we assume $n=1$ and for the parameters we let $M=-\frac{\kappa}{2}$, $Q=\frac{\sigma}{2}$ and $\Omega\Omega'=\kappa\theta$, propositions 2 and 3 yield $B(\tau,k)=\frac{\Lambda_{12}}{\Lambda_{22}}$. Now to find Λ_{12} and Λ_{22} by proposition 2 we have

$$E=\begin{pmatrix} -\frac{\kappa}{2} & -\frac{1}{2}\left(k^2+\frac{1}{4}\right) \\ -\frac{\sigma^2}{2} & -\frac{\kappa}{2} \end{pmatrix}.$$

$\Lambda=e^{\tau E}$ is a 2×2 matrix with

$$\Lambda_{12}(\tau,k)=\frac{\left(\kappa^2-\zeta^2\right)\left(-e^{\frac{\tau\zeta}{2}}+e^{\frac{-\tau\zeta}{2}}\right)}{-2\sigma^2\zeta},$$

$$\Lambda_{22}(\tau,k)=\frac{\sigma^2\left(-(\kappa+\zeta)e^{\frac{\tau\zeta}{2}}+(\kappa-\zeta)e^{\frac{-\tau\zeta}{2}}\right)}{-2\sigma^2\zeta},$$

where $\zeta=\sqrt{\kappa^2+\sigma^2\left(k^2+\frac{1}{4}\right)}$. Therefore,

$$B(\tau,k)=\frac{\Lambda_{12}(\tau,k)}{\Lambda_{12}(\tau,k)}$$
$$=\left(k^2+\frac{1}{4}\right)\frac{1-e^{-\tau\zeta}}{(\kappa+\zeta)+(-\kappa+\zeta)e^{-\tau\zeta}},$$

The function $A(\tau,k)$ can be found by integration from. This gives the price of equity call option in the presence of Heston stochastic volatility (as in [9] Equations (3.5)-(3.7)).

The price of a European put option on the equity is calculated by discounting the risk-neutral expectation of the payoff at maturity. Similarly to payoff of the call option, one can check that $\left(K-S_T 1_{\eta>T}\right)^+=\left(K-S_T\right)^+ 1_{\{\eta>T\}}+K 1_{\{\eta\le T\}}$. Therefore, the price of the put option could be rewritten as

$$
\begin{aligned}
&V_{\text{put}}(t,\Sigma,S,K) \\
&= E^{Q}_{(t,\Sigma,S)}\left(\exp\left(-\int_t^T r(s)\,\mathrm{d}s\right)\left(K - S_T 1_{\eta>T}\right)^{+}\right) \\
&= E^{Q}_{(t,\Sigma,S)}\left(\exp\left(-\int_t^T r(s)\,\mathrm{d}s\right)\left(\left(K - S_T\right)^{+} 1_{\eta>T} + K 1_{\eta\le T}\right)\right).
\end{aligned}
\tag{10}
$$

Equations (3) and (10) give the put-call parity for the equity options

$$
\begin{aligned}
&V_{\text{call}}(t,\Sigma,S,K) - V_{\text{put}}(t,\Sigma,S,K) \\
&= E^{Q}_{(t,\Sigma,S)}\left(\exp\left(-\int_t^T r(s)\,\mathrm{d}s\right)\left(\left(S_T - K\right)1_{\eta>T} - K 1_{\eta\le T}\right)\right) \\
&= E^{Q}_{(t,\Sigma,S)}\left(\exp\left(-\int_t^T r(s)\,\mathrm{d}s\right)\left(\left(S_T - K + K\right)1_{\eta>T} - K\right)\right) \\
&= V_{\text{call}}(t,\Sigma,S,0) - K\exp\left(-\int_t^T r(s)\,\mathrm{d}s\right).
\end{aligned}
$$

2.2.2. Survival Probabilities and Credit Default Swaps

Suppose $P(t,T,S_i)$ is the survival probability for the ith company

$$
P(t,T,S_i) = 1_{(t,\Sigma,S_i)}\left(S_i(\tau) > 0 \mid t < \tau \le T\right),
$$

then using Feynman-Kac formula, $P(t,T,S_i)$ satisfies the partial differential equation $P_t + \mathbf{A}_{(S,\Sigma)}P = 0$.

Proposition 4: The survival probability for the ith firm is given by

$$
P(t,T,S_i) = \frac{2e^{\frac{y}{2}}}{\pi}\int_0^{+\infty} \frac{e^{A(\tau,k)+Tr(B(\tau,k)\Sigma)}k\sin(ky)}{k^2+\frac{1}{4}}\,\mathrm{d}k,
$$

with functions $A(\tau,k)$ and $B(\tau,k)$ as in equation (7).

Credit default swaps are one of the most popular credit derivatives traded in the market. A CDS provides protection against the default of a firm, known as reference entity. The buyer of the contract pays periodic payments, called CDS spreads, until the default time or maturity date. In return, the seller of the CDS provides the buyer with the unrecovered part of the notional if default occurs. The valuation problem of a CDS is then to give the CDS spread a value such that the contract begins with a zero value. This means that the value of the floating leg and the fixed leg should coincide when the contract is written. Assume that the CDS spread is denoted by sp the periodic payments occur at $0 = T_0 < T_1 < \cdots < T_N = T$, the notional is N, the time of default is denoted by τ and the recovery rate is the constant R. The fixed leg of the CDS is the value at time $t = 0$ of the cash flow corresponding to the payments the buyer makes. With the above notation we have

$$\begin{aligned}\text{Fixed} &= E^Q\left(\sum_{i=0}^{N} e^{-\int_t^{T_i} r(s)ds} SpN\left(T_i - T_{i-1}\right)1_{\tau \geq T_i}\right) \\ &= SpN\sum_{i=0}^{N} e^{-\int_t^{T_i} r(s)ds}\left(T_i - T_{i-1}\right)Q\left(t, T_i\right).\end{aligned} \tag{11}$$

On the other hand, the floating leg, which is the value of the protection cash flow at $t = 0$, is

$$\begin{aligned}\text{Floating} &= E^Q\left(\sum_{i=0}^{N} e^{-\int_t^{T_i} r(s)ds}\left(1-R\right)1_{T_{i-1} < \tau \leq T_i}\right) \\ &= \left(1-R\right)\sum_{i=0}^{N} e^{-\int_t^{T_i} r(s)ds}\left(Q\left(t, T_{i-1}\right) - Q\left(t, T_i\right)\right)\end{aligned} \tag{12}$$

The CDS spread Sp is chosen such that the contract has a fair value at $t = 0$. By setting the fixed leg equal to the floating leg, the equations and imply

$$Sp = \frac{\left(1-R\right)\sum_{i=0}^{N} e^{-\int_t^{T_i} r(s)ds}\left(Q\left(t, T_{i-1}\right) - Q\left(t, T_i\right)\right)}{\sum_{i=0}^{N} e^{-\int_t^{T_i} r(s)ds} Q\left(t, T_i\right)}.$$

3. THE CREDITGRADES PRINCIPAL COMPONENT MODEL

In this section, we first present an stochastic eigenvalue process which is used for the covariance of the assets process. The section then covers pricing of derivatives using the CreditGrades model. We first remind the formal definition below:

Definition 2: The instantaneous stochastic covariance follows a Principal Component Model if:

$$\Sigma_t = A\Lambda_t A' \tag{13}$$

where Λ_t is a $d \times d$ diagonal matrix whose elements $\Lambda_{ii}(t) = V_i(t)$ are real valued CIR process defined, for $i = 1,2,\cdots,p$, by:

$$d\lambda_i(t) = \kappa_i(\theta_i - \lambda_i)dt + \sigma_i\sqrt{\lambda_i}dZ_i(t) \tag{14}$$

the Z_i's are independent one-dimensional Brownian motion and E is an orthogonal constant matrix. We also assume $\kappa_i > 0$, $\kappa_i\theta_i \geq \frac{\sigma_i^2}{2}$ for $i = 1,2,\cdots,p$ andwithout lost of generality, $0 < \theta_i < \theta_{i-1}$.

The main ingredient of this multivariate process is a family of one-dimensional stochastic processes for the eigenvalues. We assume for simplicity Heston-type processes but this approach works for other kind of processes.

The conditions $\kappa_i > 0$, $\kappa_i\theta_i \geq \frac{\sigma_i^2}{2}$ ensure stationarityergodicity and mixing conditions for the one-dimensional processes λ_i (see [4]). The constraints $0 < \theta_i < \theta_{i-1}$ ensure that the eigenvalues process will keep, on average, the same order but their paths could eventually cross over. This ordering on average allows us to keep the eigenvalues with greatest mean reverting levels while dropping the less significant ones.

3.1. The Dynamics of the Assets

We assume that the ith firm's value $A_i(t)$ is driven by the dynamics

$$dA_i(t) = \text{diag}(A_i(t))\left[(r(t) - d_i(t))Idt + \sqrt{\Sigma_t}dW_t\right]$$

where $W_t \in M_{m\times 1}$, $A_t \in M_{n\times 1}$, $\Sigma_t = ED_tE'$ with

$$D_t = \text{diag}(\lambda_i)_{i=1}^m \; ; E = (\alpha_{ij})_{n\times m}$$

Each λ_i follows a CIR process of the type

$$d\lambda_i(t) = \kappa_i(\theta_i - \lambda_i)dt + \sigma_i\sqrt{\lambda_i}dZ_t^i$$

In the two assets case, the above dynamics follows

$$\begin{aligned} dA_1(t) = A_1(t)\Big((r(t) - d_1(t))dt &+ \alpha_{11}\sqrt{\lambda_1(t)}dW_1(t) \\ &+ \alpha_{12}\sqrt{\lambda_2(t)}dW_2(t)\Big) \\ dA_2(t) = A_2(t)\Big((r(t) - d_2(t))dt &+ \alpha_{21}\sqrt{\lambda_1(t)}dW_1(t) \\ &+ \alpha_{22}\sqrt{\lambda_2(t)}dW_2(t)\Big) \end{aligned} \tag{15}$$

where the eigenvalues of the covariance process follow

$$\begin{aligned} d\lambda_1(t) &= \kappa_1(\theta_1 - \lambda_1)dt + \sigma_1\sqrt{\lambda_1(t)}dZ_t^1 \\ d\lambda_2(t) &= \kappa_2(\theta_2 - \lambda_2)dt + \sigma_2\sqrt{\lambda_2(t)}dZ_t^2 \end{aligned}$$

And assuming ς as the angle that the first eigenvector makes with the real axis, the eigenvector matrix E is given by

$$E = \begin{pmatrix} \cos(\varsigma) & -\sin(\varsigma) \\ \sin(\varsigma) & \sin(\varsigma) \end{pmatrix}$$

We assume that assets are driven by the Brownian motion W_t, the covariance matrix of the assets is driven by the Brownian motion Z_t and two Brownian motions W_t and Z_t are uncorrelated. The reason we make the independence assumption between stock and its volatility is that closed form formulas for the value of double-barrier options and equity options are not available when the asset and its volatility are correlated as pointed out by [8, 9].

The infinitesimal generator of the joint process (S,Σ), $\mathbf{A}_{(S,\Sigma)}$, appears in the pricing PDE. Here we find a fomula for this operator to use it for our pricing purposes in the next section. Since $S_i(t)=A_i(t)-D_i(t)$, the equity satisfies the stochastic differential equation

$$\mathrm{d}S_t=\left[\left(r_t-d_t\right)S_t\right]\mathrm{d}t+\left[S_t+D_t\right]\Sigma_t^{\frac{1}{2}}\mathrm{d}W_t$$

$\mathbf{A}_{(S,\Sigma)}$ can be divided into three terms related to the stock's operator, the covariance operator and their joint operator

$$\mathbf{A}_{(S,\Sigma)}=\mathbf{A}_S+\mathbf{A}_\Sigma+\mathbf{A}_{(S,\Sigma)}$$

Since $\mathrm{d}Z_t$ and $\mathrm{d}W_t$ are independent, the last term is zero. From the dynamics of the equity, we know that

$$\begin{aligned}
&\mathbf{A}_S(W)\\
&=\left(r(t)-d_1(t)\right)S_1W_{S_1}+\frac{1}{2}\left(S_1+D_1(t)\right)^2\left(\sum_{j=1}^{m}\alpha_{1j}^2\lambda_j\right)W_{S_1S_1}\\
&=\left(r(t)-d_2(t)\right)S_2W_{S_2}+\frac{1}{2}\left(S_2+D_2(t)\right)^2\left(\sum_{j=1}^{m}\alpha_{2j}^2\lambda_j\right)W_{S_2S_2}
\end{aligned} \tag{16}$$

And from the classical results regarding the infinitesimal generator of the CIR process

$$\mathbf{A}_\Sigma(W)=\sum_{i=1}^{m}\left[\kappa_i\left(\theta_i-\lambda_i\right)W_{\lambda_i}+\frac{1}{2}\sigma_i^2\lambda_iW_{\lambda_i\lambda_i}\right]$$

Therefore, if W is a derivative on the first underlying asset only, we have

$$\mathbf{A}_{(S,\Sigma)}(W)$$
$$=\left(r(t)-d_1(t)\right)S_1W_{S_1}+\frac{1}{2}\left(S_1+D_1(t)\right)^2\left(\sum_{j=1}^{m}\alpha_{1j}^2\lambda_j\right)W_{S_1S_1}$$
$$+\sum_{i=1}^{m}\left[\kappa_i(\theta_i-\lambda_i)W_{\lambda_i}+\frac{1}{2}\sigma_i^2\lambda_iW_{\lambda_i\lambda_i}\right]$$

In the next section, we derive closed formulas for the price of equity options and marginal probabilities of default.

3.2. Derivative Pricing; Analytical Results

In a model with two underlyings, the first asset follows the following process:

$$dA_1(t)=A_1(t)\Big(\left(r(t)-d_1(t)\right)dt+\alpha_{11}\sqrt{\lambda_1(t)}dW_1(t)$$
$$+\alpha_{12}\sqrt{\lambda_2(t)}dW_2(t)\Big)$$
$$d\lambda_1(t)=\kappa_1(\theta_1-\lambda_1)dt+\sigma_1\sqrt{\lambda_1(t)}dZ_t^1$$
$$d\lambda_2(t)=\kappa_2(\theta_2-\lambda_2)dt+\sigma_2\sqrt{\lambda_2(t)}dZ_t^2$$

We will show next the prices of several derivatives as seen from a credit perspective.

3.2.1. Equity Call Options

Calculating equity option prices is essential to calibrate the stochastic correlation CreditGrades model since this model uses the information available from the equity options to estimate the parameters of the model. Later, we will use the evolutionary algorithm method to match the theoretical results of our extended CreditGrades model with the market data. One of the advantages of the CreditGrades model compared to Merton's model, is the straight forward link it makes with the equity option markets. The price of the equity option can be calculated by discounting the payoff function at the maturity. The only subtle point here in pricing these options lies in the specific dynamics of the equity itself and the possibility of default for the company. In Black-Scholes model, the stock follows geometric Brownian motion which is a strictly positive process with a log-normal distribution and never hits zero. In the CreditGrades model, equity is modeled as a process satisfying a shifted log-normal distribution which hits the state zero when the company defaults. Because of the absorbing property of the state zero for the equity process, there is a resemblance in pricing the equity options and the pricing of the downand-out options. By considering the barrier condition for equity, the payoff of an equity call option is given by $\left(S_T 1_{\{\eta>T\}} - K\right)^+$. Therefore the price of an equity call option can be written as:

$$\begin{aligned} V_{\text{call}}(t,\Sigma,S,K) &= E^Q_{(t,\Sigma,S)}\left(\exp\left(-\int_t^T r(s)\,\mathrm{d}s\right)\left(S_T 1_{\eta>T} - K\right)^+\right) \\ &= E^Q_{(t,\Sigma,S)}\left(\exp\left(-\int_t^T r(s)\,\mathrm{d}s\right)\left(S_T - K\right)^+ 1_{\eta>T}\right) \end{aligned}$$

Similarly, the payoff of an equity put option is $\left(K - S_T 1_{\eta>T}\right)^+$. Therefore, the price of an equity put option is given by:

$$\begin{aligned} &V_{\text{put}}(t,\Sigma,S,K) \\ &= E^Q_{(t,\Sigma,S)}\left(\exp\left(-\int_t^T r(s)\,\mathrm{d}s\right)\left(K - S_T 1_{\eta>T}\right)^+\right) \\ &= E^Q_{(t,\Sigma,S)}\left(\exp\left(-\int_t^T r(s)\,\mathrm{d}s\right)\left(\left(K - S_T\right)^+ 1_{\eta>T} + K 1_{\eta\le T}\right)\right) \end{aligned}$$

Equation (18) give the put-call parity for the equity options:

$$\begin{aligned} &V_{\text{call}}(t,\Sigma,S,K) - V_{\text{put}}(t,\Sigma,S,K) \\ &= V_{\text{call}}(t,\Sigma,S,0) - K\exp\left(-\int_t^T r(s)\,\mathrm{d}s\right) \end{aligned} \tag{18}$$

The following proposition gives a closed form solution for the price of an equity call option on the first asset. Proposition C5 and equation (18) give the price of an equity put option. This result is an essential tool to calibrate the model in the next section.

Proposition 5: The price of a call option on $S_1(t)$ with maturity date T and strike price K is given by:

$$W(t,S_j)=(D_1(T)+K)\exp\left(-\int_t^T r(s)\mathrm{d}s\right)Z(\tau,y)$$

With

$$Z(\tau,y)=\mathrm{e}^y-\mathrm{e}^b-\frac{\mathrm{e}^{\frac{1}{2}y}}{\pi}$$

$$\int_0^{+\infty}\frac{\mathrm{e}^{A(\tau,k)+\sum_{i=1}^m B_i(\tau,k)\lambda_i}\left(\cos(yk)-\cos((y-2b)k)\right)}{k^2+\frac{1}{4}}\mathrm{d}s$$

$$B_i(\tau,k)=-\alpha_{1i}^2\left(k^2+\frac{1}{4}\right)\frac{1-\mathrm{e}^{-\zeta_i\tau}}{\psi_-^{(i)}+\psi_+^{(i)}\mathrm{e}^{-\zeta_i\tau}}$$

$$A(\tau,k)=\sum_{i=1}^m-\frac{\kappa_i\theta_i}{\sigma_i^2}\left(\tau\psi_+^{(i)}+2\ln\left(\frac{\psi_-^{(i)}+\psi_+^{(i)}\mathrm{e}^{-\tau\zeta_i}}{2\zeta_i}\right)\right)$$

$$y=\ln\left(\frac{S+D(t)}{D(T)+K}\right)+\int_t^T(r(s)-\mathrm{d}(s))\mathrm{d}s$$

$$b=\ln\left(\frac{D(t)}{D(T)+K}\right)+\int_t^T(r(s)-\mathrm{d}(s))\mathrm{d}s$$

$$\psi_\pm^{(i)}=\mp\kappa_i+\zeta_i$$

$$\zeta_i=\sqrt{\kappa_i^2+\alpha_{1i}^2\sigma_i^2\left(k^2+\frac{1}{4}\right)}$$

3.2.2. Survival Probabilities and Credit Default Swaps

Similar techniques can be used to find the marginal probabilities of default. Suppose $P(t,T,S_i)$ is the survival probability for the ith company

$$P(t,T,S_i)=Q_{(t,\Sigma,S_i)}\left(S_i(\tau)>0\middle|t<\tau\leq T\right)$$

Using the Feynman-Kac formula, $P(t,T,S_i)$ satisfies the partial differential equation $P_t+\mathbf{A}_{(S,\Sigma)}P=0$ with boundary conditions $P(t,T,0)=0$ and $P(T,T,S)=1$. We have the following proposition for the survival probabilities Proposition 6: The survival probability for the ith firm is given by

$$P(t,T,S_i)=\frac{2e^{\frac{y}{2}}}{\pi}\int_0^{+\infty}\frac{e^{A(\tau,k)+\sum_{j=1}^{m}B_j(\tau,k)\lambda_j}k\sin(ky)}{k^2+\frac{1}{4}}dk$$

Knowing the probability of the default, one can find the CDS spread for the underlying company. Assume that the CDS spread is denoted by S, the periodic payments occur at $0=T_0<T_1<\cdots<T_N=T$, the notional is N, the time of default is denoted by τ and the recovery rate is the constant R. The fixed leg of the CDS is the value at time $t=0$ of the cash flow corresponding to the payments the buyer makes. With the above notation we have

$$\begin{aligned}\text{Fixed}&=E^Q\left(\sum_{i=0}^{N}e^{-\int_t^{T_i}r(s)ds}SN(T_i-T_{i-1})1_{\tau\geq T_i}\right)\\&=SN\sum_{i=0}^{N}e^{-\int_t^{T_i}r(s)ds}(T_i-T_{i-1})Q(t,T_i).\end{aligned}$$

On the other hand the floating leg, which is the value of the protection cash flow at $t=0$, is

$$\text{Floating} = E^{Q}\left(\sum_{i=0}^{N} e^{-\int_t^{T_i} r(s)\mathrm{d}s}(1-R)1_{T_{i-1}<\tau\le T_i}\right)$$
$$= (1-R)\sum_{i=0}^{N} e^{-\int_t^{T_i} r(s)\mathrm{d}s}\left(Q(t,T_{i-1})-Q(t,T_i)\right)$$

The CDS spread s is chosen such that the contract has a fair value at $t=0$. By setting the fixed leg equal to the floating leg, the Equation (18) imply

$$S = \frac{(1-R)\sum_{i=0}^{N} e^{-\int_t^{T_i} r(s)\mathrm{d}s}\left(Q(t,T_{i-1})-Q(t,T_i)\right)}{\sum_{i=0}^{N} e^{-\int_t^{T_i} r(s)\mathrm{d}s} Q(t,T_i)}.$$

4. CONCLUSION

We presented a structural credit risk model which considers stochastic correlation between the assets of the companies. The stochasticity of the volatility and correlation comes from first a Wishart process and then a principal component stochastic covariance process which drives the covariance matrix of the assets. To model credit risk, we use the so called CreditGrades model. Using the affine properties of the joint log-price and volatility process, we solved the pricing problem of the equity options. We used our analytical techniques to derive quasi closed-form solution for equity options, probabilities of defaults and prices of CDSs issued by the companies.

REFERENCES

1. M.-F. Bru, "Wishart Processes," Journal of Theoretical Probability, Vol. 4, No. 4, 1991, pp. 725-751.

2. M. Escobar, B. Gotz, L. Seco and R. Zagst, "Pricing of a CDO on Stochastically Correlated Underlyings," Quantitative Finance, Vol. 10, No. 3, 2007, pp. 265-277.

3. C. Gourieroux, J. Jasiak and R. Sufana, "Derivative Pricing with Multivariate Stochastic Volatility: Application to Credit Risk," Working Paper, 2004.

4. C. Gourieroux and R. Sufana, "The Wishart Autoregressive Process of Multivariate Stochastic Volatility," Econometrics, Vol. 150, No. 2, 2009, pp. 167-181.

5. J. DaFonseca, M. Grasselli and F. Ielpo, "Estimating the Wishart Affine Stochastic Correlation Model Using the Empirical Characteristic Functionk," Working Paper ESILV, RR-35, 2007.

6. J. DaFonseca, M. Grasselli and C. Tebaldi, "A Multifactor Volatility Heston Model," Quantitative Finance, Vol. 8, No. 6, 2006, pp. 591-604.

7. J. DaFonseca, M. Grasselli and C. Tebaldi, "Option Pricing When Correlations Are Stochastic: An Analytical Framework," Review of Derivatives Research, Vol. 10, No. 2, 2007, pp. 151-180.

8. A. Lipton, "Mathematical Methods for Foreign Exchange: A Financial Engineers Approach," World Scientific, Singapore, 2001.

9. A. Sepp, "Extended Creditgrades Model with Stochastic Volatility and Jumps," Wilmott Magazine, 2006, pp. 50- 62.

10. S. L. Heston, "A Closed-Form Solution for Options with Stochastic Volatility, with Applications to Bond and Currency Options," Review of Financial Studies, Vol. 6, No. 2, 1993, pp. 327-343.

11. R. C. Merton, "On the Pricing of Corporate Debt: The Risk Structure of Interest Rates," Journal of Finance, Vol. 29, No. 2, 1974, pp. 449-470.

12. R. Stamicar and C Finger, "Incorporating Equity Derivatives into the Creditgrades Model," RiskMetrics Group, Tampa, 2005.

13. H. Abou-Kandil, G. Freiling, V. Ionescu and G. Jank, "Matrix Riccati Equations in Control and Systems Theory," Springer, Berlin, 2003.

14. M. Grasselli and C. Tebaldi, “Solvable Affine Term Structure Models,” Mathematical Finance, Vol. 18, No. 1, 2004, pp. 135-153.

CHAPTER 8

The Calibration of Some Stochastic Volatility Models Used in Mathematical Finance

Lorella Fatone[1], Francesca Mariani[2], Maria Cristina Recchioni[3], Francesco Zirilli[4]

[1] Dipartimento di Matematica e Informatica, Università di Camerino, Camerino (MC), Italy

[2] Dipartimento di Scienze Economiche, Università degli Studi di Verona, Verona, Italy

[3] Dipartimento di Management, Università Politecnica delle Marche, Ancona, Italy

[4] Dipartimento di Matematica "G. Castelnuovo", Università di Roma "La Sapienza", Roma, Italy

ABSTRACT

Stochastic volatility models are used in mathematical finance to describe the dynamics of asset prices. In these models, the asset price is modeled as a stochastic process depending on time implicitly defined by a stochastic differential Equation. The volatility of the asset price itself is modeled as a stochastic process depending on time whose dynamics is described by a stochastic differential Equation. The stochastic differential Equations for the asset price and for the volatility are coupled and together with the necessary initial conditions and correlation assumptions constitute the model. Note that the stochastic volatility is not observable in the financial markets. In order to use these models, for example, to evaluate prices of derivatives on the asset or to forecast asset prices, it is necessary to calibrate them. That is, it is necessary to estimate starting from a set of data of the values of the initial volatility and of the unknown parameters that appear in the asset price/volatility dynamic Equations. These data usually are observations of the asset prices and/or of the prices of derivatives on the asset at some known times. We analyze some stochastic volatility models summarizing merits and weaknesses of each of them. We point out that these models are examples of stochastic state space models and present the main techniques used to calibrate them. A calibration problem for the Heston model is solved using the maximum likelihood method. Some numerical experiments about the calibration of the Heston model involving synthetic and real data are presented.

KEYWORDS

Stochastic Volatility Models; Calibration

1. INTRODUCTION

Stochastic processes, stochastic differential Equations and partial differential Equations are used to describe the dynamics of random phenomena. The study of their relations in mathematical physics begins with the work of Einstein that establishes the relation between Brownian motion and the heat Equation. More recently random media and wave propagation in random media have been widely studied . In a random medium, wave motion is studied as solution of a differential Equation with random coefficients and/or random source terms. The random coefficients and the random source terms represent the medium and usually are stochastic processes with known statistical properties. In 1973, Black and Scholes introduced stochastic differential Equations in mathematical finance to model the random behavior of asset prices. The Black and Scholes asset price model can be reduced to Brownian motion with a change of variables. In particular, in the Black and Scholes model, the option pricing problem can be solved with an explicit formula (i.e. Black and Scholes formula). This model is elementary and widely used in the financial markets. However, it is known that in several circumstances the asset price dynamics model proposed by Black and Scholes in 1973 is inadequate to describe the stock prices actually observed in the financial markets. In particular, the assumption made by Black and Scholes that the asset price volatility is a constant has been criticized. To overcome this difficulty, around 1990, several stochastic volatility models of the asset price dynamics had been introduced in the theory and in the practice of mathematical finance. Models of this type are: the Hull and White model, the Stein and Stein model, the Heston model, the SABR model and many others that are not possible to mention individually. In these models the asset price and its volatility at any given time are described by random variables. The time evolution of these random variables is implicitly defined by a system of stochastic differential Equations. This system of stochastic differential Equations equipped with initial conditions and assumptions on the correlations constitutes the model. The stochastic volatility cannot be observed in the financial markets, while the asset price is observable. This means that the stochastic volatility models can be seen as stochastic state space models where the stochastic volatility Equation is the state Equation and the asset price Equation is the observation Equation. The state space models have been introduced by Kalman in 1960 in the study of guidance and control problems in aerospace engineering and are widely used. The stochastic volatility models of mathematical finance depend on parameters. Moreover, the initial stochastic volatility, which cannot be observed, must be considered as a parameter. The calibration problem consists in determining the value of these parameters from a set of data. Usually the data are observations of asset prices and/or of prices of options (derivatives) having the asset as underlying at some known times. That is, calibrating a stochastic volatility model means to solve an inverse problem for a stochastic dynamical system. The use of these models in practice is based on the solution of the calibration problem.

Several methods have been developed to solve the calibration problem. Let us mention some of them: generalized method of moments, maximum entropy, maximum likelihood, and statistical tests. This review paper introduces the stochastic volatility models, explains some of the calibration problems associated to them, illustrates the main calibration methods, and considers in detail as an example the problem of calibrating the Heston model with the maximum likelihood method using synthetic and real data.

2. BACKGROUND

In their seminal work Black and Scholes model the dynamics of an asset price through a stochastic differential Equation. Starting from the idea of Bachelier presented in his doctoral thesis at the École Polytechnique in Paris of using Brownian motion to model the uncertainty in price behavior Black and Scholes assume that the asset price evolves according to a geometric Brownian motion with constant drift and volatility. Let t be a real variable that denotes time and S_t, $t > 0$, be a stochastic process that describes the asset price as a function of time. According to the Black and Scholes model the time dynamics of S_t, $t > 0$, is governed by the following stochastic differential Equation:

$$dS_t = \mu S_t dt + \sigma S_t dW_t, t > 0 \tag{1}$$

where μ is the constant drift coefficient (i.e. expected return), σ is the constant volatility coefficient of the asset price S_t, W_t, $t > 0$, is a standard Brownian motion and dW_t, $t > 0$, is its stochastic differential.

Note that the Black and Scholes model (1) is not a stochastic volatility model. The analysis of this model produces a simple and analytically convenient formula for the option prices (the Black and Scholes formula) that has become a standard tool to evaluate European call and put options. In 1973 Merton published a paper on pricing theory enhancing the option formula derived by Black and Scholes. For their works Scholes and Merton received the Nobel Prize in Economics in 1997. Unfortunately Black died in 1995. Although the Black and Scholes model is adequate to describe the asset price dynamics in short time intervals, when we consider long time series of option prices the model does not capture some characteristic features of the volatility implied by the observed option prices known as volatility smile and skew. In particular given the observed option prices the volatilities implied by the Black and Scholes formula for options with different strike and maturity are different. This is in contrast with the assumption of the Black and Scholes model that the volatility is a constant. The time series of asset and option prices observed in the financial markets show that the volatility of the asset price is a non-constant quantity that tends to return over time to a long-term mean level, that is the behavior of the volatility can be described satisfactorily by a stochastic mean-reverting process.

3. THE STOCHASTIC VOLATILITY MODELS

The stochastic volatility models try to overcome this inadequacy of the Black and Scholes model assuming that also the volatility is modeled as a stochastic process whose dynamics is described by a stochastic differential Equation. The presence of the volatility term in the Equation that describes the dynamics of the asset price couples the differential Equations for the asset price and for its stochastic volatility generating a system of two stochastic differential Equations. This is the simplest setting of a stochastic volatility model in mathematical finance. These Equations can be interpreted as a model where the asset price propagates in a random medium described by the stochastic volatility. That is stochastic volatility models are somehow similar to the models of propagation in random media used in mathematical physics.

The first authors that have suggested the use of a stochastic volatility model in mathematical finance are Hull and White in 1987 . The Hull and White model describes the asset price dynamics as follows:

$$\begin{aligned} dS_t &= \mu S_t dt + \sqrt{v_t} S_t dW_t, \quad t > 0, \\ dv_t &= \theta v_t dt + \xi v_t dB_t, \quad t > 0. \end{aligned} \tag{2}$$

where v_t, $t > 0$, denotes the stochastic variance of the asset price S_t, $t > 0$, at time t, i.e. the volatility σ_t, $t > 0$, is given by $\sigma_t = \sqrt{v_t}, t > 0$, θ, ξ are, respectively, the drift and the volatility coefficients of the variance v_t and μ is the drift coefficient of the asset price. The quantities θ, ξ, μ are real constants and $\theta < 0$. The stochastic processes W_t, B_t, $t > 0$, are standard Brownian motions with correlation coefficient ρ, $-1 < \rho < 1$ and dW_t, dB_t, $t > 0$, are their stochastic differentials. Note that in (2) the stochastic variance is lognormally distributed and that when $\theta < 0$, we have a zero-reverting process, i.e. the variance tends to return over time to a long-term zero level. When $\rho = 0$ using a second order Taylor expansion with base point $\xi = 0$ Hull and White derived an explicit formula for the option prices whose underlying asset price dynamics is described by (2). Moreover the formula for the option prices was used only when θ is zero. The biggest weakness of this formula is that it is accurate when the variance is almost non-stochastic. In the general case (i.e. $\rho \neq 0$, $\theta \neq 0$) the option pricing problem in the Hull and White model is solved numerically. Note that the assumption $\rho = 0$ is not supported by observations of asset prices in the financial markets. In fact the observed prices show that usually ρ is negative.

The Stein and Stein model assumes that:

$$
\begin{aligned}
&dS_t = \mu S_t dt + \sigma_t S_t dW_t, \quad t > 0, \\
&d\sigma_t = \eta(\theta - \sigma_t)dt + \xi dB_t, \ t > 0,
\end{aligned}
\tag{3}
$$

where μ, η, θ, ξ are real constants such that $\eta > 0$ and W_t, B_t, $t > 0$, are standard Brownian motions with correlation ρ, $-1 < \rho < 1$.

In the Stein and Stein model when $\eta > 0$, the stochastic volatility σ_t, $t > 0$ follows a mean-reverting Ornstein-Uhlenbeck process. Note that in this model the volatility can become negative. The Stein and Stein model overcomes the difficulty of the Hull and White model to provide effective option pricing formulae providing a closed-form formula for the prices of the options on a underlying whose price dynamics is described by (3). This formula is deduced under the assumption $\rho = 0$.

A very popular stochastic volatility model is the Heston model . The Heston model assumes that the asset price S_t, $t > 0$, and its stochastic variance v_t, $t > 0$, satisfy the following Equations:

$$\begin{aligned} dS_t &= \mu S_t dt + \sqrt{v_t} S_t dW_t, \qquad t > 0, \\ dv_t &= \eta(\theta - v_t) dt + \xi \sqrt{v_t} dB_t, \; t > 0, \end{aligned} \tag{4}$$

where θ is the long-term average of v_t, $t > 0$, $\eta > 0$ is the rate of mean-reversion, ξ is the volatility of volatility coefficient. The quantities μ, θ, η, ξ are real constants, $\eta > 0$, and W_t, B_t, $t > 0$, are standard Brownian motions with correlation coefficient ρ, $-1 < \rho < 1$. In the Heston model the variance follows a Cox-Ingersoll-Ross (CIR) process . The popularity of the Heston model is due to the fact that, when $-1 < \rho < 1$, it is possible to deduce a closed form formula for the price of an European call (put) option whose underlying asset price and its stochastic volatility satisfy Equations (4). The formulae for the option prices are expressed as integrals of explicitly known integrands. These integrals must be evaluated numerically.

Note that the stochastic volatility models mentioned above assume that the asset price follows the Black and Scholes Equation (1) with stochastic volatility, the difference between them consists in the stochastic process that governs the dynamics of the volatility.

Since the pioneering work of the previous authors many other stochastic volatility models have been introduced in mathematical finance. It is not possible to mention all of them. For the convenience of the reader we just mention one of them: the SABR model . The name SABR is the acronym for "Stochastic Alpha, Beta, Rho" in fact α, β, ρ are the parameters of the model. The SABR model is widely used to price derivatives on interest rates and currency exchange rates. In the SABR model the time evolution of the asset price and of its stochastic volatility satisfy the following system of stochastic differential Equations:

$$\begin{aligned} dS_t &= \sigma_t S_t^{\beta} dW_t, \quad t > 0, \\ d\sigma_t &= \alpha \sigma_t dB_t, \quad t > 0, \end{aligned} \tag{5}$$

where α and β are real constants such that $\alpha > 0$, $0 \le \beta \le 1$ and W_t, B_t, $t > 0$, are standard Brownian motions with correlation coefficient ρ, $-1 < \rho < 1$. The parameter α is called volatility of volatility coefficient. Note that the SABR model is a stochastic version of the constant elasticity of variance (CEV) model with skewness parameter β and that it reduces to the CEV model when $\alpha = 0$. The stochastic volatility models are systems of stochastic differential Equations and to determine their solution they must be equipped with initial conditions assigned, for example, at time $t = 0$.

Unlike the asset price, in the financial markets the volatility cannot be observed. That is the stochastic volatility models are systems of stochastic differential Equations for two dependent variables, one of these variables (the asset price) is observable the other variable (the stochastic volatility or variance) is not observable. If we interpret the stochastic volatility (or variance) as a state variable and the asset price as an observation variable, the stochastic volatility models can be seen as stochastic state space models of a continuous-time dynamic system. In fact the state (or transition) Equation of the model will be the stochastic differential Equation for the volatility (or variance) and the observation (or measurement) Equation of the model will be the Equation for the asset price. Kalman for a special class of these models has developed a method to estimate the unknown parameters and the state variable from a time series of values of the observation variable. This technique is known as Kalman filter or linear quadratic estimation (LQE) algorithm. Today state space models and Kalman filters are widely used in science and engineering to study time series of data and to forecast future events based on past data.

4. CALIBRATION METHODS

The problem of identifying the parameters of stochastic volatility models starting from observed data is an inverse problem known in mathematical finance as "calibration problem". There are many calibration problems associated to stochastic volatility models depending on the data used in the calibration process. Usually the data of a calibration problem are a time series of asset prices and/or of option prices. Note that when option prices are used as data of the calibration problem it is necessary to compute their theoretical value in the stochastic volatility model considered as a function of the values assigned to the parameters. This theoretical value is defined as the expected value of the discounted payoff of the option under a risk neutral probability measure. The stochastic volatility models discussed previously admit infinitely many risk neutral measures. These measures are parameterized by a real parameter called risk premium parameter, also known as market price of volatility risk. Therefore the use of option prices as data for the calibration problem introduces an extra parameter to estimate: the risk premium parameter. Note that inverse problems for stochastic state space models are widely studied in many branches of science and engineering. However, the use of option prices as data is a feature that characterizes these problem in the context of mathematical finance.

There are several calibration methods, and we discuss briefly the following ones: the generalized method of moments, the maximum entropy method, the maximum likelihood method and the method of statistical tests.

In the generalized method of moments the unknown parameters solution of the calibration problem are those that minimize the difference between the moments defined by the model of certain variables (i.e. theoretical moments) and the same moments estimated from the data (i.e. sample moments). In order to use this method it is necessary to know the analytic expression of the (theoretical) moments or to compute them numerically. Note that the moments of the asset price variable are the natural candidates for the application of the method of moments. However other choices are possible. For example in the authors starting from the observations of the asset prices at discrete times apply the generalized method of moments to calibrate square root continuous time stochastic volatility models.

They use as theoretical moments the moments of the integrated volatility conditioned to the observations of the asset price and as sample moments the sample moments of the realized volatility.

The maximum entropy method has his origin in statistical mechanics in the study of the properties of thermodynamic systems at equilibrium. The idea of the maximum entropy principle, which is the basis of the homonymous method for the calibration of stochastic volatility models, is to find the values of the unknown parameters of the model as those that maximize the entropy associated to the probability distribution of the asset price and of its stochastic volatility conditioned to the observations. The entropy of a probability distribution measures the randomness in the probability distribution and in the context of statistical mechanics coincides with the thermodynamic entropy. Many authors have developed variants of the maximum entropy method, in particular in the optimal parameters solution of the calibration problem are found as those that minimize the relative entropy between the probability distribution associated to the model and a reference distribution.

The maximum likelihood method determines the unknown parameters of the stochastic volatility model as the parameter values that maximize a likelihood function associated to the data. The likelihood function is a function that measures the probability of obtaining the observed data as a function of the parameter values. The evaluation of the likelihood function usually requires the knowledge of the (joint) probability density function of the asset price and of its volatility. When the model used is a stochastic volatility model one of the variables (i.e. the volatility) is not observable so that the use of filtering techniques is necessary to evaluate the likelihood function. Note that given a stochastic volatility model there are several possibilities of defining a meaningful likelihood function. In a closed-form approximation of a likelihood function associated to the Heston model is derived using a series expansion. The approximate likelihood function is maximized to find the parameters of the Heston model. In 2006 Bates has found an approximation of a likelihood function associated to exponentially affine stochastic processes. This approximation is built starting from the computation of the inverse Fourier transform of the joint characteristic function of the asset price and its stochastic volatility on the observed asset prices. The parameters estimated maximizing the likelihood function are used to forecast the unknown volatility values. In the joint probability density function of the Heston model conditioned to the observations is obtained as solution of a sequence of initial value problems for the Fokker-Planck Equation of the model (i.e. as solution of the Kushner Equation). The likelihood function of the parameters can be expressed through integrals of this transition probability density function. The main feature that makes the calibration problem studied in interesting is the use as data of observations of asset prices and of option prices. The use of this set of data guarantees that the parameters obtained maximizing the likelihood function determine a good match between theoretical option prices and option prices observed in the past and produce satisfactory forecast values of asset and option prices.

Note that the maximum likelihood method is statistically more efficient than the generalized method of moments when this last method is applied using a finite number of moments. In fact the probability density function used in the likelihood function contains more "information" about the model than the "information" contained in a finite number of moments of a variable of the model. However when the evaluation of the likelihood function is difficult or computationally expensive the generalized method of moments can be a valid alternative to the maximum likelihood method.

The methods described previously provide estimates for the values of the unknown parameters of the models studied but do not associate a statistical significance level to them. The use of statistical hypothesis testing can overcome this weakness of the previous methods. In this case, the null hypothesis of a statistical test states that the values of the model parameters are equal to the values estimated with one of the methods mentioned previously. Recall that the significance level is the maximum probability of rejecting the null hypothesis when the null hypothesis is true. Starting from the data sample available of the observed asset and/or option prices, the statistical test technique can be used to test the null hypothesis with a given statistical significance level. The idea of using statistical tests to solve calibration problems for stochastic dynamical systems is a very natural one. In a statistical test procedure to calibrate the Black and Scholes model is presented. The procedure makes use of elementary statistical tests, that is: the Student's T and the χ^2 tests . The data used to calibrate the model are the observations on a discrete set of time values of the asset price. Note that when the asset price dynamics follows the Black and Scholes model the asset price log-return increments (i.e. the difference between the asset price log-returns evaluated at different times) obtained when the observation times are equally spaced time values are independent identically distributed Gaussian random variables. This implies that the calibration can be done using the elementary statistical tests already mentioned. In more general contexts when stochastic volatility models are considered it is necessary to build ad hoc statistical tests. For example in two statistical tests to calibrate the normal SABR model (with statistical significance level) in correspondence of two different set of data are presented. These statistical tests are based on some new formulae for the moments of the normal SABR model variables and make use of numerical methods.

For the convenience of the reader Table 1 summarizes the calibration methods discussed in this Section and their properties.

5. AN EXAMPLE: THE CALIBRATION OF THE HESTON MODEL

Let us consider the problem of calibrating the Heston model using the maximum likelihood method. As data of the calibration problem we use a time series of asset and option prices. This problem is an example of calibration of a stochastic volatility model and its solution shows the concrete possibilities of the calibration methods discussed in Section 4. A similar problem has been studied in .

Let us begin completing the Heston model dynamic Equation (4) with the following initial conditions:

$$S_o = \tilde{S}_o, v_o = \tilde{v}_o, \tag{6}$$

where $\tilde{S}_o, \tilde{v}_o$ are given random variables. We assume that $\tilde{S}_o, \tilde{v}_o$ are concentrated in a point with probability one, for simplicity, we identify $\tilde{S}_o, \tilde{v}_o$ with the points where they are concentrated and we assume $\tilde{S}_o, \tilde{v}_o > 0$.

Table 1. Summary of calibration methods and of their properties.

	Calibration method	Foundation of the method	Comment
1	Generalized method of moments	The unknown parameters are obtained minimizing the difference between the moments defined by the model of some random variable and the same moments estimated from the data.	This is the most popular calibration method. Option prices can be seen as moments of random variables. In this case the calibration can be reduced to the least squares fit of theoretical and observed option prices.
2	Maximum Entropy	The unknown parameters are obtained maximizing the entropy of the probability distribution of the asset price and of its stochastic volatility conditioned to the observations. The entropy is a measure of the randomness of a probability distribution.	This method is based on ideas taken from statistical mechanics. Both option prices and asset prices can be used as data. The entropy function can be substituted with an approximate entropy function to save computation.
3	Maximum Likelihood	The unknown parameters are obtained maximizing a likelihood function associated to the data. The likelihood function measures the probability of obtaining the observed data as a function of the parameter values.	Both option prices and asset prices can be used as data. The method exploits the fact that the joint probability distribution of the state variables of the models considered admits a simple integral representation formula.
4	Statistical Tests	The unknown parameters are determined using statistical hypothesis testing. Ad hoc statistical tests are developed to exploit the available data.	This method associates to the values of the parameters determined as solution of the calibration problem a statistical confidence. The ad hoc statistical tests built are based on sampling suitable random variables using numerical simulation.

Recall that in the financial markets the initial asset price $\tilde{S}_o$ can be observed while the initial stochastic variance $\tilde{v}_o$ cannot be observed and must be regarded as a parameter of the model. This means that the Heston model parameters that must be evaluated from the data in the calibration problem are:

$\mu, \eta, \theta, \xi, \rho, \tilde{v}_o$. When we consider option prices as data of the calibration problem to the previous parameters we must add the risk premium parameter λ. In fact the option prices are evaluated with respect to the risk neutral measure. This is the probability measure associated to the Heston model where η and θ are substituted respectively with $\eta^* = \eta + \lambda, \theta^* = \frac{\eta\theta}{\eta + \lambda}$. Let R^7 be the seven dimensional real Euclidean space, the unknown of the calibration problem is the vector $\underline{\Theta} = (\mu, \eta, \theta, \xi, \rho, \tilde{v}_o, \lambda)^T \in R^7$ where T denotes the transpose operator. Elementary considerations suggest that $\underline{\Theta}$ must belong to the following set:

$$M = \left\{ \underline{\Theta} = (\mu, \eta, \theta, \xi, \rho, \tilde{v}_o, \lambda)^T \in \mathbb{R}^7 \middle| \eta > 0, \xi > 0, \theta > 0, \frac{2\eta\theta}{\xi^2} \geq 1, -1 \leq \rho \leq 1, \tilde{v}_o > 0, \lambda > -\eta \right\}. \quad (7)$$

The condition $\frac{2\eta\theta}{\xi^2} \geq 1$ is imposed in order to guarantee that when $\tilde{v}_o > 0$ with probability one from Equations (4) follows that v_t is positive with probability one for $t > 0$.

The data of the calibration problem studied are:

1) The observation times $0 = t_0 < t_1 < t_2 < \cdots < t_n < +\infty$

2) The observed asset price $\tilde{S}_i$ at time $t = t_i,\ i = 0, 1, \cdots, n,$

3) The observed option price $\tilde{C}_i$ at time $t = t_i,\ i = 0, 1, \cdots, n.$

The option prices are the prices of a European call option on the asset whose price is described by the Heston model with (known) maturity time $T > t_n$ and strike price $E > 0$.

We assume that the observation times and the observed asset prices are not affected by observation errors and that the asset prices are described by the Heston model, while for $i = 1,2,\cdots,n$ the i-th observed option price when compared to the option price of the Heston model is affected by an additive observation error that we assume to be normally distributed with mean zero and known variance ϕ_i.

Let $p(S,v,t,S',v',t'),S,S',v,v',t,t' > 0,t-t' \geq 0,$ be the joint probability density function of the stochastic processes implicitely defined by (4), (6), that is let $p(S,v,t,S',v',t')$ be the probability density function of having $S_t = S, v_t = v$ given the fact that we have $S_{t'} = S', v_{t'} = v', t > t' \geq 0$. Moreover let $C(S,v,t,E,T,\underline{\Theta})$ be the price at time t when we have $S_t = S, v_t = v$ in the Heston model of parameters $\underline{\Theta}$ of a European call option with maturity time $T > t$ and strike price E. In the theoretical option price $C(S,v,t,E,T,\underline{\Theta})$ we use as discount factor μ instead of the risk free interest rate r (i.e. we assume $\mu = r$). This choice simplifies a little bit the calibration problem. The probability density function p and the call option price C can be written as one dimensional integrals of known integrand functions (see, for example). To keep the exposition simple we omit these formulae.

In the formulation of the calibration problem that follows we use the joint probability density function of the variables $S_t, v_t, t > 0$, of the Heston model conditioned to the observations $\tilde{S}_i, \tilde{C}_i$ made at the times t_i such that $t_i < t, i = 0,1,\cdots,n$. This probability density function is a piecewise solution of the forward Kolmogorov Equation associated to the Heston model. That is the transition probability density function conditioned to the observations between two consecutive observation times $t_i, t_{i+1}, i = 0,1,\cdots,n,$ (we define $t_{n+1} = +\infty$) is the solution of the forward Kolmogorov Equation that satisfies a suitable initial condition at time $t = t_i, i = 0,1,\cdots,n$. The initial condition imposed at time $t = t_i$ depends from the observations made up to time $t_i, i = 0,1,\cdots,n,$

and on the assumptions made about the observation errors (see). For $i=0,1,\cdots,n$ let $p_i(S,v,t|\underline{\Theta})$ be the joint probability density function conditioned to the observations made up to time t when $t_i < t \le t_{i+1}$. The functions $p_i, i=0,1,\cdots,n,$ can be expressed using integral representation formulae involving the transition probability function p (see). To keep the exposition simple we omit these formulae here.

Let us define the likelihood function $F(\underline{\Theta}), \underline{\Theta} \in M$, used in the formulation the calibration problem, that is:

$$F(\underline{\Theta}) = \sum_{i=0}^{n-1} \log\left[\int_0^{+\infty} dv \; p_i\left(\tilde{S}_{i+1}, v, t_{i+1} | \underline{\Theta}\right) \pi_1\left(\tilde{S}_{i+1}, v, t_{i+1} | \underline{\Theta}\right)\right] + \log\left[\int_0^{+\infty} dv \; p_0\left(\tilde{S}_0, v, t_0 | \underline{\Theta}\right) \cdot \pi_1\left(\tilde{S}_0, v, t_0 | \underline{\Theta}\right)\right], \tag{8}$$

where

$$\pi_1\left(\tilde{S}_i, v, t_i | \underline{\Theta}\right) = \frac{1}{\sqrt{2\pi\phi_i}} e^{-\frac{1}{2\phi_i^2}\left(\tilde{C}_i - C\left(\tilde{S}_i, v, t_i, E, T, \underline{\Theta}\right)\right)^2}, i=0,1,\cdots,n \tag{9}$$

It is easy to see that the choices of $\underline{\Theta} \in M$ that make big the function $F(\underline{\Theta})$ make big the probability of doing the observations $\tilde{S}_i, \tilde{C}_i$ actually done at time $t=t_i, i=0,1,\cdots,n$. This fact justifies the name likelihood function given to the function $F(\underline{\Theta}), \underline{\Theta} \in M$, defined in (8), (9). Note that the choice of the function $F(\underline{\Theta}), \underline{\Theta} \in M$, made in (8), (9) is only one possible choice of likelihood function, there are many other ways of defining legitimate likelihood functions.

The calibration problem for the Heston model is formulated as follows:

$$\max F(\underline{\Theta}). \; \underline{\Theta} \in M \tag{10}$$

This is a constrained optimization problem that must be solved numerically. The objective function of the optimization problem $F(\underline{\Theta}), \underline{\Theta} \in M$, contains several integrals that must be evaluated numerically. That is in the numerical solution of problem (10) the function evaluation can be computationally "expansive". However since the numerical evaluation of integrals is highly parallelizable when needed parallel computing can be easily used to speed up the evaluation of the function $F(\underline{\Theta}), \underline{\Theta} \in M$. The constraints contained in M are very simple and are easy to handle numerically. In the numerical experience presented below we solve the calibration problem (10) using a variable metric steepest ascent algorithm.

Let us present two numerical experiments using synthetic and real data.

Experiment 1: Calibration of the Heston model using synthetic data. We consider a "year" made of 252.5 trading days, a temporal interval of 36 consecutive trading days and n = 9 observation times in this time period.

We choose $t_0 = 0$, and $t_i = t_{i-1} + \Delta,\ i = 1, 2, \cdots, 9,$ where $\Delta = \frac{4}{252.5}$. The vector $\tilde{\underline{\Theta}}$ used to generate one trajectory of the Heston model is:

$$\underline{\Theta} = (\mu, \gamma, \varepsilon, \theta, \rho, \tilde{v}_o, \lambda)^{\mathrm{T}} = \tilde{\underline{\Theta}}$$
$$= (0.026, 5.94, 0.306, 0.01159, -0.576, 0.5, 0)^{\mathrm{T}}.$$

We have chosen $\tilde{S}_o = 1, \phi_i = 10^{-4},\ i = 0, 1, \cdots, 9$. In Table 1 we have defined $\tilde{x}_i = \log \frac{\tilde{S}_i}{\tilde{S}_o},\ i = 0, 1, \cdots, 9,$ and the notation eX means 10^{X}. Table 2 shows the data generated.

We have $F(\tilde{\underline{\Theta}}) \cong 43.15$. The initial guess of the optimization method is chosen equal to

$\underline{\Theta}^0 = (0.5, 9.22, 0.15, 0.0067, -0.634, 0.3, 0.01) \in M$. We have $\left\|\tilde{\underline{\Theta}} - \underline{\Theta}^0\right\| / \left\|\tilde{\underline{\Theta}}\right\| \approx 0.398, F(\underline{\Theta}^0) = -979.1$. Let $\underline{\Theta}^*$

be the solution of the calibration problem (10) obtained using the variable metric steepest ascent algorithm, we have $F(\underline{\Theta}^*) = 43.17$ and $\left\|\tilde{\underline{\Theta}} - \underline{\Theta}^*\right\| / \left\|\tilde{\underline{\Theta}}\right\| \approx 0.041.$

Table 2. Time series of simulated data.

i	t_i (year)	$\tilde{x}_i$	$\tilde{C}_i$
0	0	0	0.5099
1	4/252.5	1.235e−1	0.6410
2	8/252.5	6.893e−2	0.5804
3	12/252.5	−2.9809e−2	0.4791
4	16/252.5	1.2049e−2	0.5201
5	20/252.5	4.4983e−2	0.5538
6	24/252.5	−7.7662e−4	0.5065
7	28/252.5	−1.3411e−2	0.4936
8	32/252.5	−1.2461e−1	0.3894
9	36/252.5	−7.10141e−2	0.4377

Experiment 2: Analysis of the S & P 500 index. We analyze the daily closing values of the US S&P 500 index and the corresponding bid prices of a European call option on the US S&P 500 index with maturity date December 16, 2005 and strike price E = 1200 during the period of about four months going from January 3, 2005 to May 11, 2005. Due to the number of data actually available in 2005 the time unit is a year made of 253 trading days. We choose as time t = 0 the day January 3, 2005. We consider 81 windows of 10 consecutive daily observation times. In particular the observation times are chosen as $t_{i,j} = (i+j)/253,\ i = 0,1,\cdots,9, j = 0,1,\cdots,80,$ that is in each window we consider daily observations and when we move from a window to the next one we discard the observations made in the first day of the window and we add the next daily observations as last observations to build the new window. The calibration problems procedure, applied to the eighty-one windows considered provides eighty-one values of the parameter vector $\underline{\Theta}$ of the Heston model. The parameter vectors obtained from the calibration problems are substantially independent from the data window considered. Finally for $j = 0,1,\cdots,80$ we use the parameter vector obtained as solution of the calibration problem using the data of the j-th window to forecast using a filtration method (see) the value of the S&P 500

index and the bid price of the option the day next to the last observation time of the j-th window. In the eighty-one calibration and forecasting problems solved the mean relative error made in the forecasts of the next day S&P 500 index value and option bid price are 0.0063 and 0.0785 respectively.

From the results obtained in Experiment 1 and 2 we can conclude that the validity of the Heston model to interpret financial data and the quality of the calibration procedure used are confirmed.

6. FUTURE RESEARCH DIRECTIONS

The stochastic volatility models described in the previous sections can be generalized in many different ways. For example multiscale versions of stochastic volatility models are used in finance to model commodity prices or to evaluate insurance premia. These multiscale models introduce one or more extra stochastic volatility terms in the model. The multiscale version of a model usually increases the number of parameters to estimate making the calibration problem more difficult and more demanding computationally than the calibration problem for the original model. In order to overcome this difficulty it is necessary to develop ad-hoc techniques to simplify the computation or it is possible to make use of parallel computing. In and in , independently, the authors introduce a multiscale stochastic volatility model that generalizes the Heston model. In the multiscale Heston model two stochastic volatilities are considered. In an extension of the maximum likelihood method has been used to calibrate the multiscale Heston model. This calibration is based on the reduction of the dimensionality of the integrals that must be computed to evaluate the likelihood function. This reduction of the dimensionality of the integrals is obtained computing analytically some of the integrals that appear in the likelihood function. On the other hand in more general situations parallel computing and GPU computing can be used to speed up the evaluation of the likelihood function.

Note that the calibration methods described in the previous section have been applied only to special classes of stochastic volatility models. For example in and in the authors limit themselves to the calibration of exponentially affine stochastic volatility models because for these models there exists an explicit formula (involving integrals) for the joint characteristic function of asset price and its stochastic volatility. It should be interesting and useful to apply the calibration methods to the study of more general classes of models.

Finally cross fertilization between different research areas is possible. In fact, for example, the methods used to solve inverse problems for state space models outside mathematical finance, such as particle filtering, can be transferred in financial mathematics and used to calibrate stochastic volatility models.

7. CONCLUSION

The problem of calibrating stochastic volatility models is crucial for their use in practical situations. The methods described in the previous sections provide different approaches to solve this calibration problem. These methods find the unknown parameters of the model as solution of optimization problems and differ in the choice of the objective function of the optimization problem considered. The method that uses statistical tests is the only one that associates a statistical significance to the solution found. However, the choice of the hypothesis to be tested in the solution of the calibration problem may be difficult when models with several parameters (five, six parameters or more) are employed. A way to overcome this difficulty consists in using a different calibration method (i.e.: generalized method of moments, maximum entropy, …) to determine the solution of the calibration problem and in using the solution found to formulate the hypothesis to be tested. To this hypothesis is given a statistical significance using the statistical test method. We can conclude that the calibration problems for the stochastic volatility models developed in the nineties have been solved satisfactorily. The dramatic change of financial markets and of computing technologies announces the development of new, more sophisticated, stochastic volatility models whose calibration will pose new challenges. We expect that new problems and ideas will arise from considering these new models and the specific features that derive from their origin in finance.

REFERENCES

1. Einstein, A. (1905) Über die von der molekularkinetischen theorie der Wärme geforderte Bewegung von in ruhenden Flüssigkeiten suspendierten Teilchen. Annalen der Physik, 322, 549-560.

2. Keller, J.B. (1964) Wave propagation in random media. Proceedings of Symposia in Applied Mathematics, 13, 227- 246.

3. Black, F. and Scholes, M. (1973) The pricing of options and corporate liabilities. Journal of Political Economy, 81, 637-659.

4. Kalman, R.E. (1960) A new approach to linear filtering and prediction problems. Journal of Basic Engineering- Transactions of the ASME, Series D, 85, 35-45.

5. Bachelier, L. (1900) Théorie de la speculation. Annales Scientifiques de l'École Normale Supérieure, 3, 21-86.

6. Merton, R.C. (1973) Theory of rational option pricing. Bell Journal of Economics and Management Science, 4, 141- 183.

7. Hull, J. and White, A. (1987) The pricing of options on assets with stochastic volatilities. Journal of Finance, 42, 281- 300.

8. Stein, E.M. and Stein, J.C. (1991) Stock price distribution with stochastic volatility: An analytic approach. Review of Financial Studies, 4, 727-752.

9. Heston, S.L. (1993) A closed-form solution for options with stochastic volatility with applications to bond and currency options. The Review of Financial Studies, 6, 327-343.

10. Cox, J.C., Ingersoll, J.E. and Ross, S.A. (1985) A theory of the term structure of interest rates. Econometrica, 53, 385- 407.

11. Hagan, P.S., Kumar, D., Lesniewski, A.D. and Woodward, E. (2002) Managing smile risk. Wilmott Magazine, 2002, 84-108.

12. Cox, J. (1975) Notes on option pricing I: Constant elasticity of diffusions. Unpublished Draft, Stanford University.

13. Harvey, A.C., Koopman, S.J. and Shephard, N. (2004) State space and unobserved component models: Theory and applications. Cambridge University Press, Cambridge.

14. Hansen, L.P. (1982) Large sample properties of generalized method of moments estimators. Ecometrica, 50, 1029- 1054.

15. Bollerslev, T. and Zhou, H. (2002) Estimating stochastic volatility diffusion using conditional moments of integrated volatility. Journal of Econometrics, 109, 33-65.

16. Jaynes, E.T. (1957) Information theory and statistical mechanics. Physical Review, 106, 620-630.

17. Shannon, C.E. (1948) A mathematical theory of communication. Bell System Technical Journal, 27, 379-423.

18. Avellaneda, M. (1998) Minimum-relative-entropy calibration of asset pricing models. International Journal of Theoretical and Applied Finance, 1, 447-472.

19. Ait-Sahalia, Y. and Kimmel, R. (2007) Maximum likelihood estimation of stochastic volatility models. Journal of Financial Economics, 83, 413-452.

20. Bates, D.S. (2006) Maximum likelihood estimation of latent affine processes. The Review of Financial Studies, 19, 909-965.

21. Mariani, F., Pacelli, G. and Zirilli, F. (2008) Maximum likelihood estimation of the Heston stochastic volatility model using asset and

option prices: An application of nonlinear filtering theory. Optimization Letters, 2, 177-222.

22. Fatone, L., Mariani, F., Recchioni, M.C. and Zirilli, F. (2012) The use of statistical tests to calibrate the Black-Scholes asset dynamics model applied to pricing options with uncertain volatility. Journal of Probability and Statistics, 2012, Article ID: 931609.

23. Johnson, R.A. and Bhattacharyya, G.K. (2006) Statistics: Principles and methods. 5th Edition, John Wiley & Sons, New York.

24. Fatone, L., Mariani, F., Recchioni, M.C. and Zirilli, F. (2013) The use of statistical tests to calibrate the normal SABR model. Journal of Inverse and Ill-Posed Problems, 21, 59-84.

25. Christoffersen, P., Heston, S. and Jacobs, K. (2009) The shape and term structure of the index option smirk: Why multifactor stochastic volatility models work so well. Management Science, 55, 1914-1932.

26. Fatone, L., Mariani, F., Recchioni, M.C. and Zirilli, F. (2009) An explicitly solvable multi-scale stochastic volatility model: Option pricing and calibration problems. Journal of Futures Markets, 29, 862-893.

27. Mergner, S. (2009) Application of state space models in finance: An empirical analysis of the time-varying relationship between macroeconomics, fundamentals and Pan-European industry portfolios. Universitätsverlag Göttingen, Göttingen.

CHAPTER 9

Influential Observations in Stochastic Model of Divisia Index Numbers with AR(1) Errors

Syed Muhammed Aqil Burney[1], Arfa Maqsood[2]

[1] Department of AS&RM, College of Computer Science and Information Systems, Institute of Business Management Karachi, Pakistan

[2] Department of Statistics, University of Karachi, Karachi, Pakistan

ABSTRACT

We use the general form of hat matrix and DFBETA measures to detect the influential observations in order to estimate the Divisia price index number when the error structure is first order serial correlation. An example is presented with reference to price data of Pakistan. Hat values show the noteworthy findings that the corresponding weights of consumer items have large influence on the parameter estimates and are not affected by the parameter of autoregressive process AR(1). Whereas DFBETAs for Divisia index numbers depend on both the weights and autoregressive parameter.

KEYWORDS

Hat Matrix, DFBETA, Divisia Index Number, Influential Observation, Autoregressive Process

1. INTRODUCTION

A number of studies are available on the detection of leverages and influential observations in a regression model when the errors are assumed to be serially correlated with AR(1) and AR(2) processes. For instance, initially, Prais and Winsten [1] , Kadiyala [2] , Girilches and Rao [3] , Maeshiro [4], and Park and Mitchell [5] have observed the significant effect of the first observation on the parameter estimates of regression model. Influence diagnostics are developed and studied by many authors, including Belsley et al. [6] , Cook ([7] [8]), Cook and Weisberg [9] , Draper and John [10] , and Draper and Smith [11] . They examined the effect of individual observation or a set of observations on the estimation of model parameters. The problem of influence diagnostic checking for the linear regression model with autocorrelated errors has, now, received much attention of the researchers and statisticians. Puterman [12] observed that the first transformed observation in linear regression model can have a large impact on the parameter estimates. Some authors agreed that the effect of not including the first observation is not always magnificent as suggested by Cochrane and Orcutt [13] . Stemann and Trenkler [14] extended the approach of Puterman [12] to the general linear model with the first order autoregressive errors and showed the effect of the presence of a constant term on a leverage point when the correlation of the error term was large in absolute value. Barry et al. [15] extended the study of influential observations to the regression model with AR(2) errors and developed the diagnostic techniques using a hat matrix.

Much literature is on hand on the influential observations diagnostic for the regression models with continuous type of regressors. There is a need to have some techniques for finding the influential observations when the dilemma of constructing the index number is concerned. The objective of this paper is to use the analytical tools of hat matrix and DFBETA measures to identify the influential observations in estimating the Divisia price index number.

The paper is organized as follows. Section 2 introduces the Divisia index number model and the role of the initial observation in estimation of model is discussed. The relevant concepts on influence diagnostics for the underlying model are presented in Section 3. An application with reference to Pakistan price data is illustrated in Section 4 and lastly, Section 5 recapitulates the results.

2. REGRESSION MODEL

The well-known Divisia index number is formulated by the following model

$$Dp_{it} = \alpha_t + \varepsilon_{it} \tag{1}$$

where $Dp_{it} = \log p_{it} - \log p_{i,t-1} = \log\left[\frac{p_{it}}{p_{it-1}}\right]$, log price-change in period t for ith commodity, α_t conman trend in the prices of all commodities at time t (the rate of inflation), and ε_{it} is the random component. The errors are assumed to be generated from the first order autoregressive scheme, that is $\varepsilon_{it} = \phi\varepsilon_{i,t-1} + u_{it}$, where $|\phi| < 1$, and $E(u_{it}) = 0, E(u_{it}u_{jt}) = \frac{\sigma^2}{w_i}\delta_{ij}$.

This yields the error structure of model (1) as

$E(\varepsilon_{it}) = 0$ and variance-covariance

$$\gamma_k = \begin{cases} \sigma^2 / w_i\left(1-\phi^2\right), k = 0 \\ \sigma^2\phi^k / w_i\left(1-\phi^2\right), \ k > 0 \end{cases} \tag{2}$$

Defining more compactly in matrix notation as follows,

$$E(\varepsilon)=0, E(\varepsilon\varepsilon')=\sigma^2 V$$

The inverse of variance-covariance matrix can easily be decomposed using cholesky decomposition and can be written as

$$V^{-1}=Q'Q$$

where Q is a lower triangular matrix of order (nT × nT), defined as

$$Q=\left[\begin{array}{ccc|ccc|c|ccc}
\sqrt{w_1}\sqrt{1-\phi^2} & \cdots & 0 & 0 & \cdots & 0 & \cdots & 0 & \cdots & 0 \\
\vdots & \ddots & \vdots & \vdots & \ddots & \vdots & \cdots & \vdots & \ddots & \vdots \\
0 & \cdots & \sqrt{w_n}\sqrt{1-\phi^2} & 0 & \cdots & 0 & \cdots & 0 & \cdots & 0 \\
\hline
-\phi\sqrt{w_1} & \cdots & 0 & \sqrt{w_1} & \cdots & 0 & \cdots & 0 & \cdots & 0 \\
\vdots & \ddots & \vdots & \vdots & \ddots & \vdots & \cdots & \vdots & \ddots & \vdots \\
0 & \cdots & -\phi\sqrt{w_n} & 0 & \cdots & \sqrt{w_n} & \cdots & 0 & \cdots & 0 \\
\hline
\vdots & \vdots & \vdots & \vdots & \vdots & \vdots & \vdots & \vdots & \vdots & \vdots \\
\hline
0 & \cdots & 0 & 0 & \cdots & 0 & \cdots & \sqrt{w_1} & \cdots & 0 \\
\vdots & \ddots & \vdots & \vdots & \ddots & \vdots & \cdots & \vdots & \ddots & \vdots \\
0 & \cdots & 0 & 0 & \cdots & 0 & \cdots & 0 & \cdots & \sqrt{w_n}
\end{array}\right]$$

The model (1) is written in matrix form

$$D_{P(nT\times 1)} = X_{(nT\times T)}\beta_{(T\times 1)} + \varepsilon_{(nT\times 1)} \tag{3}$$

It is well known that under the above assumption, the best linear unbiased estimator (BLUE) of β in model (1) could be obtained by the generalized least square (GLS) approach as given below

$$\hat{\beta}=\left(X'V^{-1}X\right)^{-1}\left(X'V^{-1}D_P\right)$$

There are different approaches to estimate the parameter vector β, in case of unknown autoregressive parameter ϕ (see Judge et al. [16]). The value of ϕ is first estimated from the data using any of the number of suggested alternatives given in Gugarati [17] . The transformation of the vector D_P and the design matrix X to the new vector $D_P^* = \hat{Q}D_P$ and matrix $X^* = \hat{Q}X$ to obtain

$$D_P^* = \begin{bmatrix} \sqrt{1-\phi^2}\sqrt{w_i}D_{P_{i1}}\iota \\ \sqrt{w_i}D_{P_{i2}}\iota - \phi\sqrt{w_i}D_{P_{i1}}\iota \\ \vdots \\ \sqrt{w_i}D_{P_{iT}}\iota - \phi\sqrt{w_i}D_{P_{iT-1}}\iota \end{bmatrix} \text{ and } X^* = \begin{bmatrix} \sqrt{1-\phi^2}\sqrt{w_i}\iota & O & \cdots & O & O \\ -\phi\sqrt{w_i}\iota & \sqrt{w_i}\iota & \cdots & O & O \\ O & -\phi\sqrt{w_i}\iota & \cdots & O & O \\ \vdots & \vdots & \ddots & \vdots & \vdots \\ O & O & \cdots & -\phi\sqrt{w_i}\iota & \sqrt{w_i}\iota \end{bmatrix}$$

where ι and O are the vectors of one and zero respectively i.e. $\iota = [1 \;\; 1 \;\; \cdots \;\; 1]'$ and $O = [0 \;\; 0 \;\; \cdots \;\; 0]'$. We can now apply the simple ordinary least square (OLS) estimator to the transformed data to obtain estimated generalized least square (EGLS) and we have

$$\hat{\beta} = \left(X^{*\prime}X^*\right)^{-1}\left(X^{*\prime}D_P^*\right) \tag{4}$$

Substituting the results in Equation (4) provides the estimator of β, the familiar Divisia index numbers, written as

$$\hat{\alpha}_t = \sum_{i=1}^{n} w_i D_{P_{it}} \quad \text{for } t = 1,2,\cdots,T \tag{5}$$

3. INFLUENCE AND HAT MATRIX

Several measures and plots have been developed to detect the influential observations in linear regression. Hat matrix is one of the common quantity that is used in detecting the influential points when the OLS procedure for estimation of regression parameter is used. The quantity is

$$H = X(X'X)^{-1}X'$$

The hat matrix for the transformed data X^* is written as

$$H^* = X^*\left(X^{*\prime}X^*\right)^{-1}X^{*\prime} = QX(X'Q'QX)^{-1}X'Q'$$

The diagonal elements of the hat matrix, denoted by h_i or h_{ii}, are used as diagnostic technique for measuring the influence of a specific observation i on regression parameter estimates. The entries of matrix depend only on the values of design matrix X, and thus they serve as a measure of the distance of an observation from the centre of data. The large diagonal values indicate potentially large impact of the corresponding observation on regression estimates and thus considered an influential point if it satisfies the criteria that have the cut off points i.e. $h_i > 2p/n$. We obtain the hat matrix for model (1) using the transformed matrix X^* described in Section 2 as follows:

$$H^* = \begin{bmatrix} w_1 & \sqrt{w_1 w_2} & \cdots & \sqrt{w_1 w_n} \\ \sqrt{w_1 w_2} & w_2 & \cdots & \sqrt{w_2 w_n} \\ \vdots & \vdots & \ddots & \vdots \\ \sqrt{w_1 w_n} & \sqrt{w_2 w_n} & \cdots & w_n \end{bmatrix} \quad \text{for each } t = 1, 2, \cdots, T \tag{6}$$

The elements of matrix clearly show that the weights of commodities determine how much the important of particular commodity is in order to find the Divisia index number, and remain same for each time period t. They are not affected by the parameter of autoregressive process. The greater the value of weight, the more influential the commodity is, irrespective of the time period, because we are assuming the same weights over the underlying time period.

Another chief role of hat matrix in finding the significant expression to assess the effect of deleting an observation on parameter estimates and predicted values. The vector DFBETA, given by Belsley et al. [6], which denotes the difference between the estimates of the vector β with and without the ith observation i.e.;

$$DFBETA_i = \hat{\beta} - \hat{\beta}_{(i)} = \frac{\left(X^{*\prime} X^*\right)^{-1} x_i^{*\prime} e_i^*}{1 - h_i^*} \tag{7}$$

where $\hat{\beta}_{(i)}$ is the estimate of β with the ith observation excluded.

Puterman [12] studied the impact of the first observation in the constant mean model and regression through the origin model with AR(1) errors. One should see the work of Stemann and Trenkler [14] on the influence technique when considering the regression model with more than one regressors in the presence and absence of constant term. On the other hand, Barry et al. [15] extended the approach of Puterman to the influence of initial observations and subset of observations in linear regression model with AR(2) errors. Our main aim is, therefore, to obtain the influential points when we are dealing with index number model. For this purpose, we use equation (7) to find the results of DFBETA measures as follows;

$$DFBETA_{itj} = \begin{cases} \dfrac{\sqrt{w_i}}{1-w_i}\dfrac{\phi^{j-1}}{\sqrt{1-\phi^2}}e_{i1}^{*} & \text{for } j=1,2,\cdots,p, \text{ and } t=1 \\ 0 & \text{for } j=1,2,\cdots,t-1, \text{ and } t=2,3,\cdots,T \\ \dfrac{\sqrt{w_i}}{1-w_i}\phi^{j-t}e_{it}^{*} & \text{for } j=t,t+1,\cdots,p, \text{ and } t=2,3,\cdots,T \end{cases} \tag{8}$$

where p denotes the number of parameters in vector β. It is clearly seen that the DFBETA values are affected by the autoregressive coefficient of AR(1) process. For t = 1, when ϕ increases to 1 in both positive and negative direction, deleting the ith observation has a great impact on the parameter estimates. When $\phi = 0$, all the entries of DFBETA matrix become zero that might reveal having the data with no influential point. Beside this all values depend on the constant factor of ith weight embodied by the first part of expression (8).

4. AN APPLICATION

In this section, we present an application to price data for Pakistan. The data consists of 374 consumer items classified in ten groups for the period from July 2002 to June 2011. The groups are food and beverages, apparel textile and footwear, house rent, fuel and lighting, household furniture and equipment, transportation and communication, recreation and entertainment, education, cleaning laundry and per. appea, and medicare. In the first phase, we compute the parameter vector β using formula (5) which demonstrates the Divisia index number when the same weights are used through the entire period. To estimate the value of ϕ, the residuals versus observation numbers are plotted in Figure 1. The plot simply verifies the stationary scenario of time series with constant mean and constant variance.

The Durbin-Watson statistic is 1.4002, indicating the presence of autocorrelation in the residual series. The estimate of ϕ using Yule-Walker procedure is obtained as 0.299. For each time period the hat values are estimated by the weights of commodities and large quantities are shown in Table 1. The highest hat entry is parallel to house rent index, implying its importance in estimating the index numbers. Others leading items include wheat flour bag, milk fresh, and electric charges for the consumption of more than 1000 units. To investigate the effect of presence of items with large hat values, we plot the estimated Divisia index numbers based on all items and without the items house rent and wheat flour bag in Figure 2. The index number estimates excluding house rent are much different from the estimates based on all items, whereas the estimates without wheat flour bag is relatively close to the index numbers with all observations.

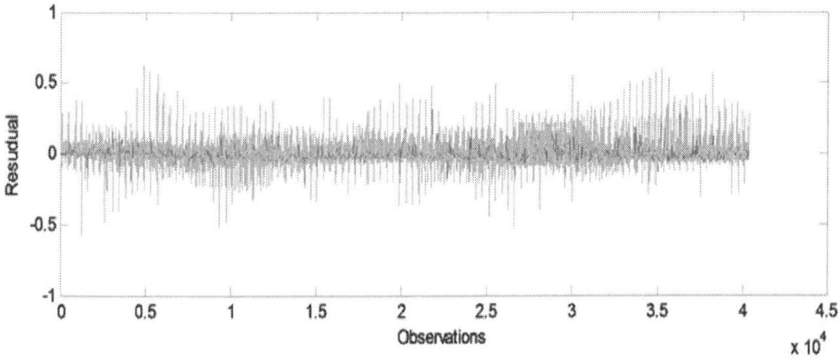

Figure 1. Plot of residual series.

Table 1. Significant hat values for the items exceeding cut-off value 0.005348.

S. Nos.	Item Nos.	Items	hi
1	2	WHEAT FLOUR FINE/SUPERIOR.	0.0134
2	3	WHEAT FLOUR BAG	0.0377
3	16	COOKING OIL (DALDA)	0.0069
4	17	VEGETABLE GHEE TIN	0.0126
5	18	VEGETABLE GHEE (LOOSE)	0.0141
6	19	SUGAR REFINED	0.0195
7	25	TEA LOOSE KENYA AV.QLTY 250G	0.0057
8	26	MILK FRESH (UNBOILED)	0.0653
9	70	TOFFEE (HILAL)	0.0101
10	71	CHOWKELATE CANDY (SMALL SIZE)	0.0099
11	86	BEEF WITH BONE AV.QLTY.	0.0161
12	87	MUTTON AV.QLTY.	0.0109
13	88	CHICKEN FARM BROILER (LIVE)	0.0092
14	91	ONION	0.0058
15	167	HOUSE RENT INDEX	0.2343
16	176	ELECT.CHARGES 301 - 1000 UNI	0.0106
17	177	ELECT.CHARGES ABOVE 1000 UNI	0.0274
18	180	GAS CHRG 6.7438 - 10.1157MMB	0.0093
19	181	GAS CHRG10.1157 - 13.4876MMB	0.0068
20	206	HOUSEHOLD SERVANT FEMALE P/T	0.0119
21	227	PETROL SUPER	0.0173
22	266	TELEPHONE CHARGES LOCAL CALL	0.0083
23	268	TEL CHARGES OUT SIDE CITY	0.0083
24	286	SCHOOL FEE PRIMARY ENG.MED.	0.0081
25	287	SCHOOL FEE 2ND-RY ENG.MED.	0.0081
26	310	WASHING SOAP NYL(135-160GMS)	0.0054
27	333	HAIRCUT CHARGES FOR MEN	0.0067
28	374	DOCTOR (MBBS) CLINIC FEE	0.01

Table 2 presents list of items with the significant DFBETAs exceeding the cutoff value 0.00995. The large values of h_i are represented by the values with superscript *. Out of these 37 influential items, 32 are from the first commodity group food and beverages that include wheat flour, vegetable ghee, sugar, beef with bones, and mutton with the corresponding higher diagonal values of hat matrix. Sugar refined with $h_{19} = 0.019467$ has an impact on estimation index number relating to 41 months or in other words, it affects the values of 41 alphas in parameter vector β. Other influential commodity groups include house rent with the highest $h_{167} = 0.234298$, fuel and lighting, and transportation and communication.

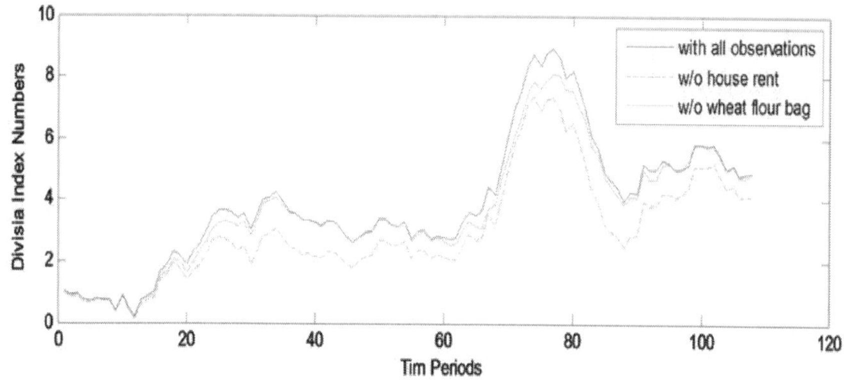

Figure 2. Divisia index numbers including all items, excluding house rent and wheat flour.

Table 2. Significant DFBETAs corresponding to items and their hat values.

S. Nos.	Item No.	Items	Hat values h_i	No. of significant DFBETAs	Significant DFBETAs for estimating α_t for $t =$
1	1	WHEAT	0.00483	6	67,71,72,73,74,76
2	2	WHEAT FLOUR FINE/SUPERIOR.	0.013398*	13	23,66,67,68,69,70,71,72,73,74,75,76,77
3	3	WHEAT FLOUR BAG	0.037724*	23	18,20,21,23,24,25,26,27,64,65,66,67,70, 71,72,73,74,75,76,77,78,80,81
4	6	RICE BASMATI SUP. QLTY.	0.003711	7	65,66,67,71,72,73,74
5	7	RICE BASMATI 385/386	0.003865	11	61,62,63,64,65,66,67,71,72,73,74
6	8	RICE BASMATI BROKEN AV.QLTY	0.004766	12	61,62,63,64,65,66,67,68,71,72,73,74
7	9	RICE IRRI-6 (SINDH/PUNJAB)	0.001027	1	71
8	10	PULSE MASOOR (WASHED)	0.002214	9	71,72,73,74,75,76,77,78,79
9	11	PULSE MOONG (WASHED)	0.00223	11	46,93,94,95,96,97,98,99,100,101,102
10	12	PULSE MASH (WASHED)	0.002017	4	47,94,95,96
11	13	PULSE GRAM	0.004272	1	55
12	16	COOKING OIL (DALDA)	0.006858*	2	69,70
13	17	VEGETABLE GHEE TIN	0.012545*	10	7,8,9,65,68,69,70,71,72,73
14	18	VEGETABLE GHEE (LOOSE)	0.014127*	24	58,59,60,61,62,63,64,65,66,67,68,69,70,71,72, 73,74,75,103,104,105,106,107,108
15	19	SUGAR REFINED	0.019467*	41	31,32,33,34,35,36,37,38,39,40,41,42,44,45,46, 47,48,79,80,81,82,83,84,85,86,87,88,89,90,91, 92,93,94,95,96,97,98,99,100,101,102
16	56	CHILLIES POWD. NATIONAL 200GM	0.003379	8	67,72,103,104,105,106,107,108

5. CONCLUSION

In this paper, we use the general expression of hat matrix and DFBETA measure to detect the influential observations in order to estimate the Divisia price index number when the error is generated from AR(1) process. The hat values only depend on the weights of commodities, showing that the corresponding weights of consumer items have large influence on the parameter estimates and are not affected by the parameter of autoregressive process AR(1). An example is presented with reference to price data of Pakistan. From the findings of both hat matrix and DFBETAs, food and beverages are the leading commodity group as the maximum number of items in which group has large hat values and DFBETAs measures.

ACKNOWLEDGEMENTS

The authors are thankful to Dept. of Computer Science and Dept. of Statistics University of Karachi for providing computing and research facilities and Dr Ejaz Ahmed, Dean CCSIS ,Institute of Business Management Karachi for technical discussion and support.

REFERENCES

1. Prais, G.J. and Winsten, C.B. (1954) Trend Estimates and Serial Correlation. Cowles Commission Discussion Paper, Stat. No. 383, University of Chicago.

2. Kadiyala, K.R. (1968) A Transformation Used to Circumvent the Problem of Autocorrelation. Econometrica, 36, 93-96.

3. Griliches, Z. and Rao, P. (1969) Small-Sample Properties of Several Two-Stage Regression Methods in the Context of Autocorrelated Disturbances. Journal of American Statistical Association, 64, 253-272.

4. Maeshiro, A. (1979) On the Retention of the First Observation in Serial Correlation Adjustment of Regression Models. International Economic Review, 20, 259-265.

5. Park, R.E. and Mitchell, B.M. (1980) Estimating the Autocorrelated Error Model with Trended Data. Journal of Econometrics, 13, 185-201.

6. Belsley, P.A., Kuh, E. and Welsch, R.E. (1980) Regression Diagnostics. John Wiley, New York.

7. Cook, R.D. (1977) Detection of Influential Observations in Linear Regression. Technometrics, 19, 15-18.

8. Cook, R.D. (1979) Influential Observations in Linear Regression. Journal of American Statistical Association, 74, 169- 174.

9. Cook, R.D. and Weisberg, S. (1982) Residuals and Influence in Regression. Chapman and Hall, New York.

10. Draper, N.R. and John, J.A. (1981) Influential Observations and Outliers in Regression. Technometrics, 23, 21-26.

11. Draper, N.R. and Smith, H. (1998) Applied Regression Analysis. 3rd Edition, John Wiley, New York.

12. Puterman, M.L. (1988) Leverage and Influence in Autocorrelated Regression Model. Journal of the Royal Statistical Society, 37, 76-86.

13. Cochrane, D. and Orcutt, G.H. (1949) Application of Least Squares Regression to Relationships Containing Autocorrelated Error Terms. Journal of American Statistical Association, 44, 32-61.

14. Stemann, D. and Trenkler, G. (1993) Leverages and Cochrane-Orcutt Estimation in Linear Regression. Communication in Statistics-Theory and Methods, 22, 1315-1333.

15. Barry, A.M., Burney, S.M.A. and Bhatti, M.I. (1997) Optimum Influence of Initial Observatins in regression Models with AR (2) Errors. Applied Mathematics and Computations, 82, 57-65.

16. Judge, G.G., Griffiths, W.E., Hill, R.C., Lutkepohl, H. and Lee, T.C. (1985) The Theory and Practice of Econometrics. 2nd Edition, John Wiley, New York.

17. Gujarati, D.N. (2003) Basic Econometrics. 4th Edition, McGraw-Hill, New York.

CHAPTER 10

Asymptotic proportion of arbitrage points in fractional binary markets

Fernando Cordero[a], Irene Klein[b], Lavinia Perez-Ostafe[c]

[a] Faculty of Technology, University of Bielefeld, Universitätsstr. 25, 33615 Bielefeld, Germany

[b] Department of Statistics and Operations Research, University of Vienna, Oskar-Morgensternplatz 1, A-1090 Vienna, Austria

[c] Department of Mathematics, ETH Zurich, Rämistrasse 101, 8092 Zurich, Switzerland

ABSTRACT

A fractional binary market is a binary model approximation for the fractional Black–Scholes model, which Sottinen constructed with the help of a Donsker-type theorem. In a binary market the non-arbitrage condition is expressed as a family of conditions on the nodes of a binary tree. We call "arbitrage points" the nodes which do not satisfy such a condition and "arbitrage paths" the paths which cross at least one arbitrage point. In this work, we provide an in-depth analysis of the asymptotic proportion of arbitrage points and arbitrage paths. Our results are obtained by studying an appropriate rescaled disturbed random walk.

KEYWORDS

Fractional Brownian motion; Fractional binary markets; Binary markets; Arbitrage opportunities

1. INTRODUCTION

In the classical theory of mathematical finance a crucial role is played by the notion of*arbitrage*, which is the cornerstone of the option pricing theory that goes back to F. Black, R. Merton and M. Scholes [2]. In the case of binary markets, the absence of arbitrage is completely characterized by Dzhaparidze in [6]. Intuitively, a binary market is a market in which the stock price process $(S_n)_{n=0}^{N}$ is an adapted stochastic process with strictly positive values and such that at time n the stock price evolves from S_{n-1} to either $\alpha_n S_{n-1}$ or $\beta_n S_{n-1}$, where $\beta_n<\alpha_n$.

One advantage of working with binary markets is given, on one hand, by their simplicity and, on the other hand, by their flexibility to approximate more complicated models. In particular, this is possible for Black–Scholes type markets that are driven by a process, for which we dispose of a random walk approximation. Examples of this are the fractional Brownian motion and the Rosenblatt process, as one can see in [14] and [15]respectively.

In this paper we provide an in-depth analysis of fractional binary markets, which are defined by Sottinen [14] as a sequence of binary models approximating the fractional Black–Scholes model, i.e. a Black–Scholes type model where the randomness of the risky asset comes from a fractional Brownian motion. Along this work we assume that the Hurst parameter H is strictly bigger than 1/2. In this case, the fractional Brownian motion exhibits self-similarity and long-range dependence, properties that were observed in some empirical studies of financial time series (see [3] and [17]). Since the fractional Brownian motion fails to be a semimartingale, the fractional Black–Scholes model admits arbitrage opportunities, a drawback that can be corrected if, e.g. one introduces transaction costs.

In [14] Sottinen constructs the fractional binary markets by giving an analogue of the Donsker theorem, where the fractional Brownian motion is approximated in distribution by a “disturbed” random walk. Sottinen proves that the arbitrage opportunities do not only appear in the limiting model, but also in the sequence of fractional binary markets.

According to [6], in a binary market, the absence of arbitrage can be written as a family of conditions on the nodes of a binary tree. We call an “arbitrage point” a node in the binary tree which does not satisfy the corresponding non-arbitrage condition. An “arbitrage path” is a path that crosses at least one arbitrage point. By [14] we know that, for each fractional binary market in the sequence, the associated set of arbitrage points is not empty.

The study of the set of arbitrage points provides a way to quantify arbitrage, a research direction which goes a step further than the classical question of existence of arbitrage.

The aim of this paper is to study qualitative and quantitative properties of the sets of arbitrage points and paths for the fractional binary market. First, we prove that starting from any point in the binary tree we reach an arbitrage point by going enough times only up or only down (Proposition 3.2). This generalizes the result of Sottinen, who showed the existence of arbitrage starting only from the root of the tree. This gives information about the structure of the set of arbitrage points and implies that its cardinality is asymptotically infinite. Next, we study the limit behaviour of the proportion of arbitrage points. The latter is expressed in terms of a rescaled random walk, which we show converges in law. The characterization of the asymptotic proportion of arbitrage points then follows (Theorem 3.3). We also take a closer look to the previous limit when H tends to 1/2 and when H tends to 1 (Proposition 3.4). Finally, making use of the 0–1Kolmogorov law, we show that when H is close to 1, a.s. a path in the binary tree crosses an infinite number of arbitrage points, and when H is close to 1/2, a.s. a path in the binary tree crosses an infinite number of non-arbitrage points (Theorem 3.5).

We believe that our asymptotic results open a way to a better understanding of the arbitrage behaviour in the limiting market. Since the proportion of arbitrage points remains strictly positive in the limit, one could expect that the sequence of sets of arbitrage points converges in a proper way to a set encoding the arbitrage structure of the fractional Black–Scholes model.

Another possible direction, in which our results may turn useful is the study of arbitrage in the fractional binary markets under transaction costs. As mentioned by Sottinen, one may expect that the arbitrage disappears when transaction costs are taken into account. This latter problem was treated in its most generality in [4], where a characterization of the smallest transaction cost (called "critical" transaction costs) starting from which the arbitrage is eliminated is provided. However, since the parameters of the model depend on time and space, this characterization does not give a closed-form solution, but reduces to solving an optimization problem in a binary tree. The complexity of this problem increases with the number of arbitrage points, and, hence, the understanding of qualitative and quantitative properties gives us an insight to this more complicated problem.

The paper is organized as follows. In Section 2, we recall the fractional binary markets as defined in [14]. In Section 3, we introduce the notion of arbitrage point and arbitrage path and we state our main results: Proposition 3.2, Theorem 3.3, Proposition 3.4 andTheorem 3.5. The remaining of the paper is devoted to their proofs. In Section 4, we prove that the parameters of the fractional binary markets satisfy a scaling property. This helps us to get rid of the dependence on the size of the fractional binary market. We finish this section with the proof of Proposition 3.2. In Section 5, we relate the proportion of arbitrage points with a rescaled random walk and study its limit behaviour. Finally, we prove Theorem 3.3 and Proposition 3.4. Section 6 contains the proof of our last main result, Theorem 3.5, concerning the asymptotic proportion of arbitrage paths. We end the paper with Appendix enclosing the most technical auxiliary results needed along this work.

2. FRACTIONAL BINARY MARKETS

Sottinen introduces in [14] the fractional binary markets as a sequence of binary markets approximating the fractional Black–Scholes model. The latter is a Black–Scholes type model where the randomness of the risky asset is determined by a fractional Brownian motion. More precisely, the dynamics of the bond and stock are given by:

$$dB_t = r(t)\, B_t\, dt \quad \text{and} \quad dS_t^H = (a(t)dt + \sigma\, dZ_t^H)\, S_t^H, \quad t \in [0,1], \tag{2.1}$$

where $\sigma > 0$ is a constant representing the volatility and Z^H is a fractional Brownian motion of Hurst parameter $H \in (1/2, 1)$. In this case, the increments of the fractional Brownian motion are positively correlated and exhibit long-range dependence, see e.g. [11] and [12]. The functions r and a are deterministic and represent the interest rate and the drift of the stock, respectively. We assume moreover that the interest rate is constant equal to 0 and that the drift a is continuous.

For each N>1, we introduce the following market, called N-period fractional binary market, which converges to the fractional Black–Scholes model (2.1) as shown in [14]. Let (Ω, F, P) be a finite probability space. The bond and the stock are traded at the times $\{0, \frac{1}{N}, \ldots, \frac{N-1}{N}, 1\}$ under the dynamics:

$$B_n^{(N)} = 1 \quad \text{and} \quad S_n^{(N,H)} = \left(1 + a_n^{(N)} + X_n^{(N,H)}\right) S_{n-1}^{(N,H)}, \quad n \geq 1.$$

The initial values are $B_0^{(N)} = 1$ and $S_0^{(N,H)} = s_0$, with s_0 a positive constant. Here, $B_n^{(N)}$ and $S_n^{(N,H)}$ are understood to be the value of $B^{(N)}$ and $S^{(N,H)}$ in the time interval $[\frac{n}{N}, \frac{n+1}{N})$ for each $n \in \{0, \ldots, N-1\}$. The drift $a^{(N)}$ is used to approximate the continuous drift given in (2.1)via

$$a_n^{(N)} = \frac{1}{N} a(n/N)$$

and therefore, for all N,

$$|a_n^{(N)}| \leq \frac{\|a\|_\infty}{N}, \quad n \in \{1, \ldots, N\}. \tag{2.2}$$

Using the construction of Sottinen, we can express $X_n^{(N,H)}$ as:

$$X_n^{(N,H)} = \sum_{i=1}^{n-1} J_n^{(N,H)}(i)\,\xi_i + g_n^{(N,H)}\,\xi_n. \tag{2.3}$$

The random variables $(\xi_i)_{i\geq 1}$ are supposed to be i.i.d. Bernoulli, i.e. $P(\xi_1=-1)=P(\xi_1=1)=1/2$. The real numbers $J_n^{(N,H)}(i)$, for $1\leq i<n\leq N$, and $g_n^{(N,H)}$, for $1\leq n\leq N$, are defined below. We endow the probability space with the following filtration $F_n=\sigma(\xi_1,\ldots,\xi_n)$, for $n\geq 1$, and $F_0=\{\emptyset,\Omega\}$, and therefore the process $S_n^{(N,H)}$ is adapted.

Now, we define the constants:

$$J_n^{(N,H)}(i) := \sigma\sqrt{N}\int_{\frac{i-1}{N}}^{\frac{i}{N}}\left(k_H\left(\frac{n}{N},u\right) - k_H\left(\frac{n-1}{N},u\right)\right)du, \tag{2.4}$$

and

$$g_n^{(N,H)} := \sigma\sqrt{N}\int_{\frac{n-1}{N}}^{\frac{n}{N}} k_H\left(\frac{n}{N},u\right)du, \tag{2.5}$$

where

$$k_H(t,s) := c_H\left(H-\frac{1}{2}\right)s^{\frac{1}{2}-H}\int_s^t u^{H-\frac{1}{2}}(u-s)^{H-\frac{3}{2}}du, \tag{2.6}$$

and $c_H := \sqrt{\frac{2H\,\Gamma\left(\frac{3}{2}-H\right)}{\Gamma\left(H+\frac{1}{2}\right)\Gamma(2-2H)}}$ is a normalizing constant. For simplicity, we will use from time to time the notation $C_H := c_H\left(H-\frac{1}{2}\right)$. We also define, for each $n\in\{1,\ldots,N\}$, the functions $Y_n^{(N,H)} : \{-1,1\}^{n-1}\to\mathbb{R}$ by:

$$Y_n^{(N,H)}(x_1,\ldots,x_{n-1}) := \sum_{i=1}^{n-1} J_n^{(N,H)}(i)\,x_i, \quad Y_1^{(N,H)} = 0.$$

We shortly denote $Y_n^{(N,H)}$ the random variable $Y_n^{(N,H)}(\xi_1,\ldots,\xi_{n-1})$. In particular, from (2.3)we have the following identity:

$$X_n^{(N,H)} = Y_n^{(N,H)} + g_n^{(N,H)} \xi_n,$$

where the first process denotes the contribution of the past (up to time n−1) and the second one depends only on the present (at time n). The following functions on $\{-1,1\}^{n-1}$ are introduced:

$$u_n^{(N)}(x_1, \ldots, x_{n-1}) := Y_n^{(N,H)}(x_1, \ldots, x_{n-1}) + g_n^{(N,H)},$$

and

$$d_n^{(N)}(x_1, \ldots, x_{n-1}) := Y_n^{(N,H)}(x_1, \ldots, x_{n-1}) - g_n^{(N,H)},$$

with $u_1^{(N)}$ and $d_1^{(N)}$ constants. Thus, given the history up to time n−1, the process $X^{(N,H)}$ can take at each time n only two possible values $u_n^{(N)}$ and $d_n^{(N)}$ with $d_n^{(N)} < u_n^{(N)}$. This justifies the binary structure of these markets.

3. ARBITRAGE POINTS AND MAIN RESULTS

In this section we introduce the notions of arbitrage points and arbitrage paths and formulate our main results concerning their asymptotic properties.

We know from [6] (see [5] for the binomial case) that the N-period fractional binary market excludes arbitrage opportunities if and only if for all $n \in \{1,\ldots,N\}$ and $x \in \{-1,1\}^{n-1}$, we have:

$$d_n^{(N)}(x) < -a_n^{(N)} < u_n^{(N)}(x). \tag{3.1}$$

The previous characterization of the arbitrage opportunities in the fractional binary market motivates the next definitions. We call the following set N-binary tree:

$$\mathbb{X}_N = \{\tau\} \cup \bigcup_{n=1}^{N-1} \{-1, 1\}^n,$$

where τ denotes the root of the tree. We say that a point $x \in X_N$ is an arbitrage point for the N-period fractional binary market if x does not satisfy condition (3.1). More precisely, given a level $n \in \{1,\ldots,N\}$, the following object is called the set of arbitrage points at level n:

$$\mathcal{A}_n^{(N,H)} := \{x \in \{-1,1\}^{n-1} : u_n^{(N)}(x) \leq -a_n^{(N)} \text{ or } d_n^{(N)}(x) \geq -a_n^{(N)}\}, \quad n \geq 2,$$

and $\mathcal{A}_1^{(N,H)}$ is equal to $\{\tau\}$ if $u_1^{(N)} \leq -a_1^{(N)}$ or $d_1^{(N)} \geq -a_1^{(N)}$ and the empty set otherwise. The set of arbitrage points is given by:

$$\mathcal{A}^{(N,H)} := \bigcup_{n=1}^{N} \mathcal{A}_n^{(N,H)} \subseteq \mathbb{X}_N.$$

In addition, we call arbitrage paths the paths in the N-binary tree which cross at least one arbitrage point, i.e. the elements of the set:

$$\mathcal{A}_{\mathcal{P}}^{(N,H)} := \{(x_1,\ldots,x_{N-1}) \in \{-1,1\}^{N-1} : \exists n \in \{1,\ldots,N\}, (x_1,\ldots,x_{n-1}) \in \mathcal{A}_n^{(N,H)}\}.$$

Remark 3.1.

Sottinen proves in [14] that for N large enough, the N-period fractional binary market admits arbitrage opportunities. Indeed, it is proved that there exists $n_H > 0$ such that for all N sufficiently large:

$$\mathcal{A}_{n_H}^{(N,H)} \neq \emptyset \quad \text{and} \quad \lim_{N\to\infty} \frac{|\mathcal{A}_{\mathcal{P}}^{(N,H)}|}{2^N} \geq 2^{2-n_H} > 0.$$

Now, we formulate our first main result. In [14, Theorem 5], the author shows that starting from the root of the binary tree and going always up we can always reach an arbitrage point. The following proposition provides a generalization of that result, establishing that starting from any point in the binary tree by going always up (or always down) we can always reach an arbitrage point. In what follows we use the notation $1_k := (1,\ldots,1) \in R^k$.

Proposition 3.2.

For all $k \geq 2$ *and* $x \in \{-1,1\}^{k-1}$, *there exist* $n_k(x) \geq 1$ *and* $N_k(x) \geq k + n_k(x)$ *such that for all* $N \geq N_k(x)$:

$$\left(x, 1_{n_k(x)}\right) \in \mathcal{A}_{k+n_k(x)}^{(N,H)} \quad \textit{and} \quad \left(x, -1_{n_k(x)}\right) \in \mathcal{A}_{k+n_k(x)}^{(N,H)}.$$

In particular,

$$\lim_{N \to \infty} | \mathcal{A}^{(N,H)} | = \infty.$$

Before stating the next main result, we introduce the following notations. For $h := \frac{H}{2} + \frac{1}{4} \in (\frac{1}{2}, \frac{3}{4})$, consider the random variable

$$\mathcal{Y}_H := 2 g_H \sum_{k=1}^{\infty} \rho_h(k)\, \xi_k, \tag{3.2}$$

where $g_H := \frac{\sigma c_H}{H + \frac{1}{2}}$ and

$$\rho_h(k) := \frac{1}{2}\left((k+1)^{2h} + (k-1)^{2h} - 2k^{2h}\right). \tag{3.3}$$

Therefore

$$\rho_h(k) \sim h(2h-1)\, k^{2h-2} = \frac{1}{8}(2H+1)(2H-1)\, k^{H-\frac{3}{2}} \quad \text{as } k \to \infty.$$

Consequently, $\sum_{k=1}^{\infty} \rho_h(k) = \infty$ and, hence, the series in (3.2) is not absolutely convergent. However, since $\sum_{k=1}^{\infty} \rho_h^2(k) < \infty$, Y_H is well defined in the sense of almost sure convergence (see p. 113 in [16] or [8]). We remark that ρ_h is the autocovariance function of a fractional Brownian motion with Hurst parameter h.

Our second main result characterizes the proportion of arbitrage points.

Theorem 3.3.

For any sequence $N_n \geq n$:

$$\lim_{n\to\infty} \frac{|A_n^{(N_n,H)}|}{2^{n-1}} = P(|\mathcal{Y}_H| > g_H) > 0.$$

In particular:

$$\lim_{N\to\infty} \frac{|A^{(N,H)}|}{2^N - 1} = P(|\mathcal{Y}_H| > g_H).$$

The next proposition provides the behaviour of the previous asymptotic proportion whenH is close to 1/2 and when H is close to 1.

Proposition 3.4.

We have that:

$$\lim_{H\to\frac{1}{2}} P(|\mathcal{Y}_H| > g_H) = 0 \quad and \quad \liminf_{H\to 1} P(|\mathcal{Y}_H| > g_H) \geq \frac{1}{3}.$$

In order to state our last result concerning the asymptotic proportion of arbitrage paths, we recall the following notions. For any sequence of measurable sets $A_1, A_2, \ldots$, we denote $\{A_n \text{ i.o.}\}$ and $\{A_n \text{ ult.}\}$, respectively the sets where A_n happens infinitely often and where A_n happens ultimately, by:

$$\{A_n \text{ i.o.}\} := \bigcap_{n\geq 1}\bigcup_{k\geq n} A_k \quad \text{and} \quad \{A_n \text{ult.}\} := \bigcup_{n\geq 1}\bigcap_{k\geq n} A_k.$$

Theorem 3.5.

There exists $H_c \in (\frac{1}{2}, 1)$ *such that for* $H>H_c$

$$P(|\mathcal{Y}_n^H| > g_n^H \text{ i.o.}) = 1$$

and for $H<H_c$

$$P(|\mathcal{Y}_n^H| > g_n^H \text{ ult.}) = 0.$$

In particular, if H>H$_c$ *then*

$$\lim_{N\to\infty} \frac{\mathcal{A}_{\mathcal{P}}^{(N,H)}}{2^{N-1}} = 1.$$

Remark 3.6.

The previous theorem tells us that, when H>H$_c$, with probability 1 a path in the binary tree asymptotically crosses an infinite number of arbitrage points. On the other hand, when H<H$_c$, with probability 1 a path in the binary tree asymptotically crosses an infinite number of non-arbitrage points.

Proof.

Fix H<H$_c$. By Corollary A.5, we can choose ε such that

$$\sum_{k=1}^{\infty} \rho_{\bar{h}}^2(k) < \frac{(1-\varepsilon)^2}{4} < \frac{1}{4}.$$

Using Fatou's lemma and Tchebysheff's inequality, we obtain:

$$P(|\widehat{\mathcal{Y}}_{n_k}^H| > g_{n_k}^H(1-\varepsilon) \text{ ult.}) \le \liminf_{k\to\infty} P(|\widehat{\mathcal{Y}}_{n_k}^H| > g_{n_k}^H(1-\varepsilon)) \le \frac{4}{(1-\varepsilon)^2}\sum_{k=1}^{\infty} \rho_{\bar{h}}^2(k) < 1.$$

Thus, the first result is a consequence of Lemma 6.1.

Now, fix H>H$_c$ and choose δ>0 such that

$$\sum_{k=1}^{\infty} \rho_{\bar{h}}^2(k) > \frac{(1+\delta)^2}{4} > \frac{1}{4}.$$

Another application of Fatou's lemma, implies that:

$$P(|\widehat{\mathcal{Y}}_{n_k}^H| > g_{n_k}^H(1+\delta) \text{ i.o.}) \ge \limsup_{n\to\infty} P(|\widehat{\mathcal{Y}}_{n_k}^H| > g_{n_k}^H(1+\delta)).$$

Applying the Paley–Zygmund inequality and Khintchine's inequality, we get:

$$\limsup_{n\to\infty} P(|\widehat{\mathcal{Y}}_{n_k}^H| > g_{n_k}^H(1+\delta)) = P(|\mathcal{Y}_H| > g_H(1+\delta))$$

$$\geq \frac{1}{3}\left(1-\left(\frac{(1+\delta)^2}{4\sum_{k=1}^{\infty}\rho_h^2(k)}\right)\right)^2 > 0,$$

and the second statement follows from Lemma 6.1. □

Proof of Theorem 3.5.

Since $\bar{\mathcal{Y}}_n^H$ converges to 0 in L^2, there is a subsequence $\bar{\mathcal{Y}}_{n_k}^H$ convergent almost surely to 0. Let $H>H_c$ and $\delta>0$ as in Lemma 6.2. Note that

$$\begin{aligned}\{|\widehat{\mathcal{Y}}_{n_k}^H| > (1+\delta)g_{n_k}^H\} &= \{|\widehat{\mathcal{Y}}_{n_k}^H| > (1+\delta)g_{n_k}^H\} \cap \{|\bar{\mathcal{Y}}_{n_k}^H| \leq \delta g_{n_k}^H\}\\ &\cup \{|\widehat{\mathcal{Y}}_{n_k}^H| > (1+\delta)g_{n_k}^H\} \cap \{|\bar{\mathcal{Y}}_{n_k}^H| > \delta g_{n_k}^H\}\\ &\subseteq \{|\mathcal{Y}_{n_k}^H| > g_{n_k}^H\} \cup \{|\bar{\mathcal{Y}}_{n_k}^H| > \delta g_{n_k}^H\},\end{aligned}$$

which implies that

$$\{|\widehat{\mathcal{Y}}_{n_k}^H| > (1+\delta)g_{n_k}^H \text{ i.o.}\} \subseteq \{|\mathcal{Y}_{n_k}^H| > g_{n_k}^H \text{ i.o.}\} \cup \{|\bar{\mathcal{Y}}_{n_k}^H| > \delta g_{n_k}^H \text{ i.o.}\}.$$

Since $\bar{\mathcal{Y}}_{n_k}^H \xrightarrow{\text{a.s.}} 0$ and $g_{n_k}^H \to g_H > 0$, it follows that $P(\{|\bar{\mathcal{Y}}_{n_k}^H| > \delta g_{n_k}^H \text{ i.o.}\}) = 0$. The desired statement is a consequence of Lemma 6.2 and of the fact that

$$\{|\mathcal{Y}_{n_k}^H| > g_{n_k}^H \text{ i.o.}\} \subseteq \{|\mathcal{Y}_n^H| > g_n \text{ i.o.}\}.$$

For $H<H_c$, let $\varepsilon>0$ as in Lemma 6.2. Note that

$$\begin{aligned}\{|\mathcal{Y}_{n_k}^H| > g_{n_k}^H\} &= \{|\mathcal{Y}_{n_k}^H| > g_{n_k}^H\} \cap \{|\bar{\mathcal{Y}}_{n_k}^H| \leq \varepsilon g_{n_k}^H\}\\ &\cup \{|\mathcal{Y}_{n_k}^H| > g_{n_k}^H\} \cap \{|\bar{\mathcal{Y}}_{n_k}^H| > \varepsilon g_{n_k}^H\}\\ &\subseteq \{|\widehat{\mathcal{Y}}_{n_k}^H| > (1-\varepsilon)g_{n_k}^H\} \cup \{|\bar{\mathcal{Y}}_{n_k}^H| > \varepsilon g_{n_k}^H\},\end{aligned}$$

which implies that

$$\{|\mathcal{Y}_{n_k}^H| > g_{n_k}^H \text{ ult.}\} \subseteq \{|\widehat{\mathcal{Y}}_{n_k}^H| > (1-\varepsilon)g_{n_k}^H \text{ ult.}\} \cup \{|\bar{\mathcal{Y}}_{n_k}^H| > \varepsilon g_{n_k}^H \text{ i.o.}\}.$$

Following the same reasoning as before, the desired statement is obtained usingLemma 6.2 and the fact that

$$\{|\mathcal{Y}_{n_k}^H| > g_{n_k}^H \text{ ult.}\} \supseteq \{|\mathcal{Y}_n^H| > g_n \text{ ult.}\}.$$

For the final statement, note that

$$\{|\mathcal{Y}_n^H| > g_n^H \text{ i.o.}\} \subseteq \{\exists\, n : |\mathcal{Y}_n^H| > g_n^H\}$$

and

$$\{\exists\, n : |\mathcal{Y}_n^H| > g_n^H\} \subseteq \{\exists\, n \in \{1, \ldots, N\} : |\mathcal{Y}_n^H + a_n^{(N)} N^H| > g_n^H \text{ ult.}\}.$$

Using (5.2), Fatou's lemma and Theorem 3.5 we obtain

$$\liminf_{N\to\infty} \frac{\mathcal{A}_P^{(N,H)}}{2^{N-1}} \geq 1$$

and the result follows.

ACKNOWLEDGEMENT

The third author gratefully acknowledges financial support from the Austrian Science Fund (FWF): J3453-N25.

REFERENCES

1. S.M. Berman, Local times and sample function properties of stationary Gaussian processes, Trans. Amer. Math.Soc. 137 (1969) 277–299.

2. F. Black, M. Scholes, The pricing of options and corporate liabilities, J. Polit. Econ. 81 (3) (1973) 637.

3. R. Cont, Long range dependence in financial markets, in: Fractals in Engineering, Springer, 2005, pp. 159–180.

4. F. Cordero, I. Klein, L. Perez-Ostafe, Binary markets under transaction costs, Int. J. Theor. Appl. Finance 17 (05)(2014) 1450030.

5. J.C. Cox, S.A. Ross, M. Rubinstein, Option pricing: A simplified approach, J. Financ. Econ. 7 (3) (1979) 229–263.

6. K.O. Dzhaparidze, Introduction to Option Pricing in a Securities Market, in: CWI Syllabi, vol. 47, Stichting Mathematisch Centrum Centrum voor Wiskunde en Informatica, Amsterdam, 2000.

7. L. H¨ormander, A uniqueness theorem of Beurling for Fourier transform pairs, Ark. Mat. 29 (2) (1991) 237–240.

8. M. Kac, Statistical Independence in Probability, Analysis and Number Theory, in: The Carus Mathematical Monographs, vol. 12, Published by the Mathematical Association of America. Distributed by JohnWiley and Sons, Inc., New York, 1959.

9. O. Kallenberg, Foundations of Modern Probability, second ed., in: Probability and its Applications (New York), Springer-Verlag, New York, 2002.

10. A. Khintchine, U¨ ber dyadische Bru¨che, Math. Z. 18 (1) (1923) 109–116.

11. B.B. Mandelbrot, J.W. Van Ness, Fractional Brownian motions, fractional noises and applications, SIAM Rev. 10 (1968) 422–437.

12. I. Norros, E. Valkeila, J. Virtamo, An elementary approach to a Girsanov formula and other analytical results on fractional Brownian motions, Bernoulli 5 (4) (1999) 571–587.

13. S.I. Resnick, A Probability Path, Birkh¨auser Boston Inc., Boston, MA, 1999.

14. T. Sottinen, Fractional Brownian motion, random walks and binary market models, Finance Stoch. 5 (3) (2001) 343–355.

15. S. Torres, C.A. Tudor, Donsker type theorem for the Rosenblatt process and a binary market model, Stoch. Anal. Appl. 27 (3) (2009) 555–573.

16. D. Williams, Probability with Martingales, in: Cambridge Mathematical Textbooks, Cambridge University Press, Cambridge, 1991.

17. W.Willinger, M.S. Taqqu, V. Teverovsky, Stock market prices and long-range dependence, Finance Stoch. 3 (1999) 1–13.

Index